Lecture Notes in Computer Science 9424

Commenced Publication in 1973
Founding and Former Series Editors:
Gerhard Goos, Juris Hartmanis, and Jan van Leeuwen

W0080000

More information about this series at http://www.springer.com/series/7409

Sebastian Gottwalt · Lukas König
Hartmut Schmeck (Eds.)

Energy Informatics

4th D-A-CH Conference, EI 2015
Karlsruhe, Germany, November 12–13, 2015
Proceedings

 Springer

Editors
Sebastian Gottwalt
FZI Forschungszentrum Informatik
Karlsruhe
Germany

Lukas König
Institut AIFB
KIT
Karlsruhe
Germany

Hartmut Schmeck
Institut AIFB
KIT
Karlsruhe
Germany

ISSN 0302-9743   ISSN 1611-3349   (electronic)
Lecture Notes in Computer Science
ISBN 978-3-319-25875-1   ISBN 978-3-319-25876-8   (eBook)
DOI 10.1007/978-3-319-25876-8

Library of Congress Control Number: 2015952764

LNCS Sublibrary: SL3 – Information Systems and Applications, incl. Internet/Web, and HCI

Printed on acid-free paper

Springer International Publishing AG Switzerland is part of Springer Science+Business Media
(www.springer.com)

Preface

Welcome to the proceedings of the 4th D-A-CH conference Energy Informatics 2015, which took place in Karlsruhe, Germany, November 12–13, 2015.

Germany, Austria, and Switzerland have set ambitious targets for increasing energy efficiency, reducing greenhouse gas emission, and enlarging the share of renewable energy sources. Energy informatics is developing the IT-based solutions required for achieving these targets. Applications include the construction of intelligent energy networks (smart grids), their utilization for integrating renewable energy sources and energy storage, increased flexibility of demand as well as the design of systems for improving total efficiency, system stability, and security of supply.

The conference series was started as an initiative of the German, Austrian, and Swiss Cooperation on Smart Grids (www.smartgrids-dach.eu). The first Energy Informatics conference took place in Oldenburg, Germany, in July 2012, followed by conferences in Vienna, Austria, in November 2013, and in Zurich, Switzerland, in November 2014.

The objective of Energy Informatics 2015 was the support of a research-based development and implementation phase of adequate information and communication technologies and to foster the transfer between academia, industry, and service providers, addressing scientists and practitioners as well.

The call for papers resulted in 36 submissions by authors from the D-A-CH region and from Denmark, Italy, and the USA. The Program Committee selected 18 papers for oral presentation and inclusion in the proceedings, using a double-blinded reviewing process with three reviews for every submission. In addition, seven submissions were suggested for presentation as a poster. This resulted in an attractive program, including two keynote talks from academia by Michael Sonnenschein (Carl-von-Ossietzky-Universität Oldenburg) and from industry by Holger Krawinkel (MVV, Mannheim), as well as another poster session presented by the participants of the 6th PhD Workshop Energy Informatics, which was held before the conference in the FZI Research Center for Information Technology at Karlsruhe.

As another highlight, the second day of the conference was supplemented by an industrial track organized by the network fokus.energie.

Sincere thanks go to all those involved actively in the organization of D-A-CH Energy Informatics 2015: the Steering Committee for their thoughtful guidance, the members of the Program Committee and the additional reviewers for selecting the program, the members of the Industry Track Committee for organizing the additional industry workshop, and to Daniel Pathmaperuma, Sebastian Gottwalt, and Lukas König for contributing essentially to the local organization.

Furthermore, special thanks go to the German Federal Ministry for Economic Affairs and Energy for the financial support and to ABB for sponsoring the conference.

November 2015   Hartmut Schmeck

Organization

Program Committee

Göran Andersson	ETH Zürich, Switzerland
H.-Jürgen Appelrath	OFFIS, Germany
Jörg Benze	T-Systems, Germany
Wilfried Elmenreich	University of Klagenfurt, Austria
Elgar Fleisch	ETH Zürich, Switzerland
Wolfgang Gawlik	TU Wien, Austria
Veit Hagenmeyer	Karlsruhe Institute of Technology, Germany
Lorenz Hilty	University of Zürich, Switzerland
Kai Hufendiek	University of Stuttgart, Germany
Gabriela Hug	ETH Zürich, Switzerland
Hans Arno Jacobsen	TU München, Germany
Andrea Kollmann	Energieinstitut Linz, Austria
Friederich Kupzog	Austrian Institute of Technology, Austria
Jean-Yves Le Boudec	EPFL, Switzerland
Sebastian Lehnhoff	OFFIS, Germany
Friedemann Mattern	ETH Zürich, Switzerland
Christoph Mayer	OFFIS, Germany
Peter Palensky	University of Delft, The Netherlands
Christian Rehtanz	University of Dortmund, Germany
Hartmut Schmeck	Karlsruhe Institute of Technology, Germany
Hans-Peter Schwefel	FTW Vienna, Austria
Michael Sonnenschein	Carl von Ossietzky Universität Oldenburg, Germany
Thorsten Staake	University of Bamberg, Germany
Anke Weidlich	Offenburg University of Applied Sciences, Germany
Christof Weinhardt	Karlsruhe Institute of Technology, Germany

Additional Reviewers

Bao, Kaibin	Kleiminger, Wilhelm	Rahmani, Mohsen
Beckel, Christian	Kozlovskiy, Ilya	Rosinger, Sven
Becker, Birger	Kuhnapfel, Uwe	Salah, Florian
Bessler, Sandford	Künzel, Thomas	Schloegl, Florian
Bremer, Jörg	Levesque, Martin	Schuller, Alexander
Bytschkow, Denis	Lüers, Bengt	Sodenkamp, Mariya
Dauer, David	Mikut, Ralf	Ströhle, Philipp
Düpmeier, Clemens	Monacchi, Andrea	Stucky, Karl-Uwe
Egarter, Dominik	Moser, Simon	Tarish, Haider
Guo, Junyao	Muggenhumer, Gerold	Vogel, Ute
Gärttner, Johannes	Müller-Syring, Gert	Wenig, Jürgen
Hinrichs, Christian	Pöchacker, Manfred	Zhang, Xiao

Invited Talks

Decentralized Provision of Active Power by Means of Dynamic Virtual Power Plants

Michael Sonnenschein

Carl von Ossietzky Universität Oldenburg
PO Box 2503, 26111 Oldenburg, Germany
sonnenschein@informatik.uni-oldenburg.de

Abstract. Currently virtual power plants are statically structured coalitions of medium sized power suppliers (and possibly controllable loads and storage systems) controlled by a centralized control unit. In this talk our model of dynamic virtual power plants (DVPPs) controlled by distributed, agent-based methods will be introduced. DVPPs are dynamically formed to provide specific power products. Compared to static VPPs they offer more flexibility to integrate small distributed units like micro-CHPs of different types, and they allow distributed units to hide some private information. Within this context, the objective of this talk is to introduce a seamless process chain for day-ahead based active power provision by means of DVPPs. In the project cluster Smart Nord we developed a multi-agent system realizing the aggregation algorithm, the reactive scheduling heuristic as well as the flexibility modelling used for DVPP management and control.

Prosumers and Disruptive Technologies
Challenges of the Energiewende

Holger Krawinkel

MVV Energie AG
Luisenring 49, 68159 Mannheim, Germany
holger.krawinkel@mvv.de

Abstract. Looking at the ongoing changes in the energy sector, the question arises of whether and how this transformation can be successful. Basically, two important principles of the energy sector are threatened: The supply and the high costs for storing electricity. Over decades electric utilities built up know-how in managing electricity flows. This knowledge, which was strengthened through the Energiewende, might lose its value soon. If storage costs should be cut in half till the end of 2016 and PV panels keep getting cheaper, the tipping point for energy autarky might be reached earlier than expected. In addition, PV might soon be integrated in the facades of buildings in near future. These changes require transferring the aggregated know-how of the electric utilities into the (energy) management of buildings.

Contents

Distributed Energy Sources
and Storage

SWARM - Increasing Households' Internal PV Consumption and Offering Primary Control Power with Distributed Batteries

David Steber[✉], Peter Bazan, and Reinhard German

Chair of Computer Networks and Communication Systems,
Friedrich-Alexander-University Erlangen-Nürnberg,
Martensstr. 3, 91058 Erlangen, Germany
{david.steber,peter.bazan,reinhard.german}@fau.de

Abstract. This *Research in Progress Paper* deals with a simulation app-roach for a virtual mass storage composed of small distributed battery energy storage units, installed in households with a roof-top photovoltaic system. On the one hand the household's internal consumption of pho-tovoltaic energy is maximized and on the other hand primary control reserve power is provided by a central storage controller. This concept is academically approved and rolled out in the field within this project. First simulation results show a household's benefit of installing a bat-tery energy storage system and an accurate working of the implemented virtual mass storage.

Keywords: Central controlled distributed batteries · Internal PV con-sumption · Primary control reserve power

1 Introduction

The German government motivates the energy transition and initialized a pro-found power system change by setting ambitious climate goals. This requires new innovative solutions, e.g. for the future provision of ancillary services, to ensure a safe and stable electricity supply system operation. Currently, these services are mainly provided by conventional power plants whose operating hours will sig-nificantly decrease in the future due to seen economic effects (e.g., Merit-Order Effect). Therefore, they will no longer be able to make their today's substan-tial contribution to ancillary services [5]. This evokes a demand for sustainable alternative solutions, as an economical and reliable system service supply is indis-pensable. Using battery energy storage systems (BESS) is one way to provide ancillary services in future due to their technical characteristics and benefit for the system. Furthermore BESS contribute to the further integration of renewable energy sources, which is also supported by the governments [13].

Grid-connected BESS are able to afford an efficient contribution to the secu-rity preservation of the electricity system [3]. Their suitability and economic efficiency for ancillary services has already been confirmed by realized projects

© Springer International Publishing Switzerland 2015
S. Gottwalt et al. (Eds.): EI 2015, LNCS 9424, pp. 3–11, 2015.
DOI: 10.1007/978-3-319-25876-8_1

in the field (e.g., in Zurich, Switzerland or Schwerin, Germany). Based on their possible steep power gradients, BESS are well-suited for the provision of ancillary services that require fast response times [11]. Furthermore, the provision of Primary Control Reserve (PCR) power based on units with a limited energy/power supply and capacity – such as distributed BESS – is confirmed to be as reliable as conventional PCR power provision [4]. By linking the buffering of photovoltaic (PV) converted energy and the supply of other services, the profitability of a single BESS can be doubled, depending on the market and the quantity of request [9]. BESS thus functionally support system's stability and further market integration of renewable energies [7]. One way to do this is to provide grid relief by reducing the PV feed-in power during times of a high solar penetration, whereby the highest benefit is obtained under the consideration of forecasting PV power plus loads [10,12]. One disadvantage of using batteries in households is the currently high investment cost. That way it is motivated using them for providing multiple services on different system levels (e.g., ancillary services) in order to increase their profitability.

Within the research project *SWARM* (Storage With Amply Redundant Megawatt) 65 BESS with a single gross capacity of 21 kWh (net cap. 18 kWh) are installed in residential houses owning a roof-top PV system. On the one hand, a BESS maximizes the households' internal consumption of provided photovoltaic power by making it available during hours with low or no solar penetration. On the other hand, all installed BESS are interconnected via mobile communication to a virtual mass BESS that can offer 1 MW PCR power and is operated by a central control algorithm.Additionally, this decentralized storage system allows to generate additional business cases for grid stabilization depending on its power, capacity and installation location.

The project *SWARM* is a cooperation between *N-ERGIE Aktiengesellschaft* and *Caterva GmbH* and funded by the Bavarian state. The 65 BESS are provided and installed until summer 2015 and also operated by *Caterva GmbH*. The 1 MW virtual storage is located in the distribution grid of *Main-Donau Netzgesellschaft mbH*. The project's research activities are further supported by groups of the *Friedrich-Alexander-University (FAU)*. That way, expertise of electricity grid calculations plus simulation and modeling of energy systems supports gaining answers on fundamental technical and economical questions, that come up with the innovative design of the virtual BESS-based mass storage. They mainly deal with the efficient and grid supporting BESS operation and the benefit from the household's point of view. Furthermore, the PCR power provision with distributed storage will be validated.

2   Operational Framework

There are many influences on BESS's operation, set-up, and profitability. From the household's point of view, different regulations (e.g., the *Renewable Energies Act of Germany (EEG)* or the *Energy Act (EnWG)*) have to be considered. That includes fees, taxes, and compensations, plus the electricity supply price, which all have a high influence on the households profitability of installing a BESS. On the local distribution grid level, connection terms and corresponding laws/guidelines have to be taken into account. This is for example important regarding the transferable power rates for providing primary control power and the BEES's configuration. Furthermore, the European transmission system operator's guidelines for primary control power provision have to be considered.

2.1 Household's Internal PV Consumption

Figure 1 shows the integration of a BESS into a household. Without the BESS, PV power can either be consumed directly or fed into the grid (light blue lines in Fig. 1). For every provided kWh of PV energy to the grid, a feed-in compensation is payed to the household based on the EEG. In addition to the saved energy supply costs for every internally consumed kWh, a fee is payed depending on the size and the start-up date of the PV system. Furthermore, the EEG reallocation charge has to be payed for every internally consumed kWh for PV systems with more than 10 kWp [6]. Installing battery storage capacity ensures a decrease of feed-in and an increase of internal consumption by shifting the surplus PV energy into times of insufficient PV feed-in (dark blue lines in Fig. 1). In conclusion, the household's benefit of either consume or feed in a provided kWh from the PV system depends on the electricity price and EEG-based fees. Furthermore, investment cost and future payments have to be taken into account for an investment decision.

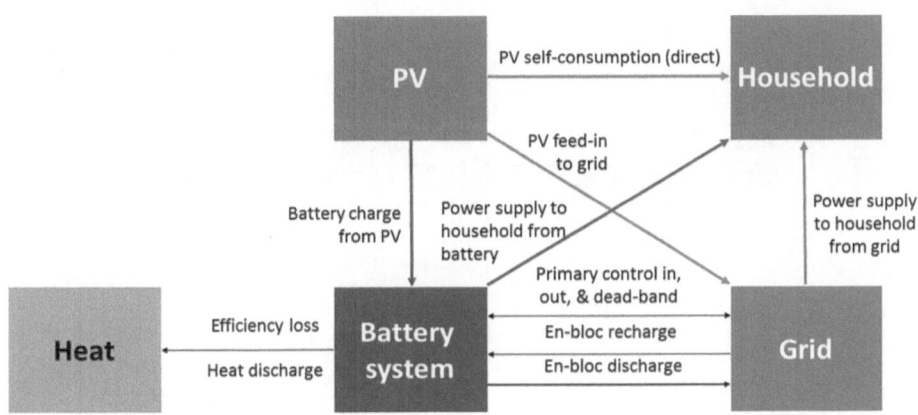

Fig. 1. Integration of a BESS into the household (Color figure online)

2.2 Primary Control Power

For PCR market participation at least 1 MW PCR power has to be provided and a qualification process has to be completed. Depending on the frequency deviation from 50 Hz, a certain amount of power has either to be taken from or provided to the grid by the virtual mass storage as expressed by Eq. 1. An optional 20 % over-fulfillment of requested PCR power is accepted, which supports a flexible mass storage's operation. Within a frequency deviation of 0.01 Hz from the rated frequency of 50 Hz (deathband), PCR power provision is only allowed in conformity with the system in order to avoid contra-productive effects [1]. Figure 2 shows an extraction of the frequency time series used in the simulations including the deathband. The PCR power provision is low, as there is in most times only a slight derivation from the rated frequency.

$$PCR\,[kW] = \begin{cases} PCR_{contracted} & \forall\, f < 49.8\,Hz, \\ PCR_{contracted} \cdot \frac{50\,Hz-f}{0.2\,Hz} & \forall\, 49.8\,Hz \leq f \leq 50.2\,Hz, \\ -PCR_{contracted} & \forall\, f > 50.2\,Hz. \end{cases} \quad (1)$$

Another point to consider is the duration (30 s), until 100 % of the automatically requested PCR power has to be available. Due to their high power gradients, BESS are said to be able to provide PCR power faster than it is recommended. That is why the *SWARM* storage is supposed to provide PCR power more efficiently and flexibly than conventional power plants. Clarifying this is one research activity within the project.

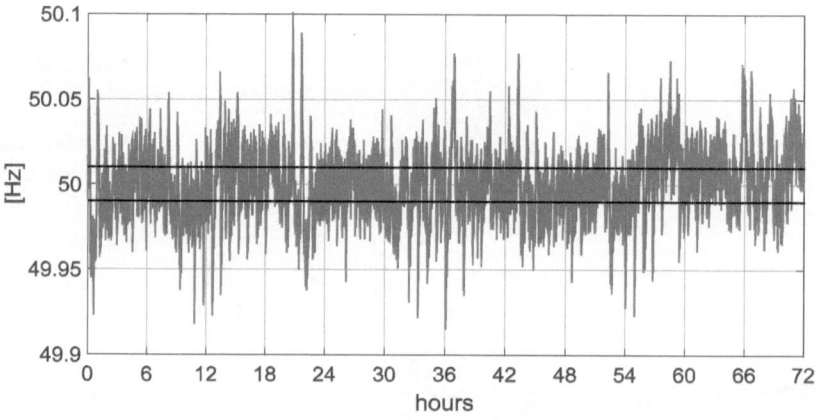

Fig. 2. Extraction of frequency time series used in simulations

3   Simulation Model

The simulation model is implemented by using the toolbox *i7-AnyEnergy* [2] in *Anylogic*. It offers a flexible and efficient framework for the hybrid simulation of

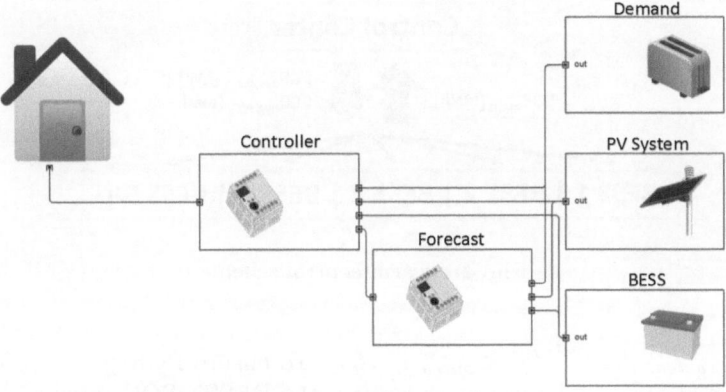

Fig. 3. *i7-AnyEnergy* house model

energy systems combining discrete event simulation (e.g., models of consumers, weather conditions, controller) and system dynamic models (e.g., energy and cost flows). A house model consists of a PV system, a BESS, a load profile generator, a forecast service, and a house controller (Fig. 3). Furthermore, there is a controller for operating the virtual mass storage and a weather model for providing solar radiation time series to the houses on an upper model level.

The load profile generation is based on stochastic variation of a standard load curve [8]. The latter is accessible for the controller as a load forecast. The forecast of the PV feed-in works similar, based on a daily solar radiation profile and its stochastic variation. The predicted household's residual load curve results from putting both forecasts together. From that, the PV energy that probably has to be stored can be derived, what makes it possible to decide whether to store PV power or to feed it into the grid. Therefore, the BESS's State of Charge (SOC) and capacity has to be taken into account at each time step. If the available BESS's capacity ($E_{cap} - E_{SOC}$) is smaller than the expected energy to be stored over the day, the PV power is partly stored and partly fed into the grid. If the expected surplus energy is less or equal the available BESS's capacity, the whole PV power is stored. That strategy avoids high power gradients of PV feed-in to the grid as the BESS is not running full before the evening so that there is no hard switch from no to 100 % PV feed-in. This paper looks also at the effect of this strategy's application and shows the influence on the household's benefit compared to a greedy strategy, where surplus PV power is stored without any restrictions until the BESS is full. If there is a high PV feed-in into the grid next to no time, this greedy algorithm can cause high power gradients that can bring about critical conditions.

Figure 4 shows the fundamental simulation model setup. Each of the 65 BESS operates on the lower level. On the top level runs the centralized control algorithm for ensuring the PCR power provision. Within this paper, it is assumed that every BESS has to deliver the same amount of PCR power

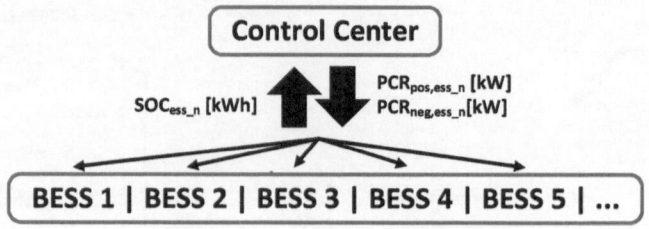

Fig. 4. Swarm control scheme

$(PCR_{contracted,BESS} = PCR_{contracted}/65)$. To ensure PCR power provision at every time (100 % availability mandatory), the BESS's SOC range for charging with PV power or discharging for internal consumption is bounded with a lower and an upper bound referred to the BESS' capacity (see SOC plots in Fig. 5). That way, it is always possible to overcome the worst case of 15 min ongoing provision of 1 MW PCR power.

4 Results

The *SWARM* project's first step focuses on identifying the household's benefit of installing a BESS. The *i7-AnyEnergy SWARM* simulation model calculates all energy flows shown in Fig. 1 separately. So it is easy to value them monetarily.

The simulation results shown in Table 1 refer to a household with an annual consumption of 3500 kWh, where a 5.75 kWp PV system is installed in February 2015. This ends up in a benefit of 0.1658 €/kWh for every internally consumed kWh based on 0.2911 €/kWh electricity price (increasing with 3 % p.a.) and 0.1253 €/kWh EEG fee. All given values are average values over 20 years BESS run time and 65 houses. The greedy (forecast) control strategy scenario has an annual PV generation of 5347 kWh (5369 kWh). Applying the greedy (forecast) control strategy and valuing all energy flows monetarily, installing a BESS means earning 84.32 € (80.34 €) instead of paying 408.65 € (406.60 €) in average every year, what results in a benefit of 492.97 € (486.94 €).

Table 1. Household's benefit simulation results with and without BESS

	Greedy strategy			Forecast strategy		
	w/o	w/	diff	w/o	w/	diff
Purchase [kWh]	2307	531	**-1776**	2309	556	**-1753**
Purchase [€]		929.08	213.64	**-715.44**	930.04	223.42	**-706.62**
int. cons. [kWh]	1193	2969	**1776**	1191	2944	**1753**
int. cons. rate [%]	22.31	55.53	**32.56**	22.18	54.83	**32.56**
Self-suff. rate [%]	34.09	84.83	**50.74**	34.03	84.11	**50.08**

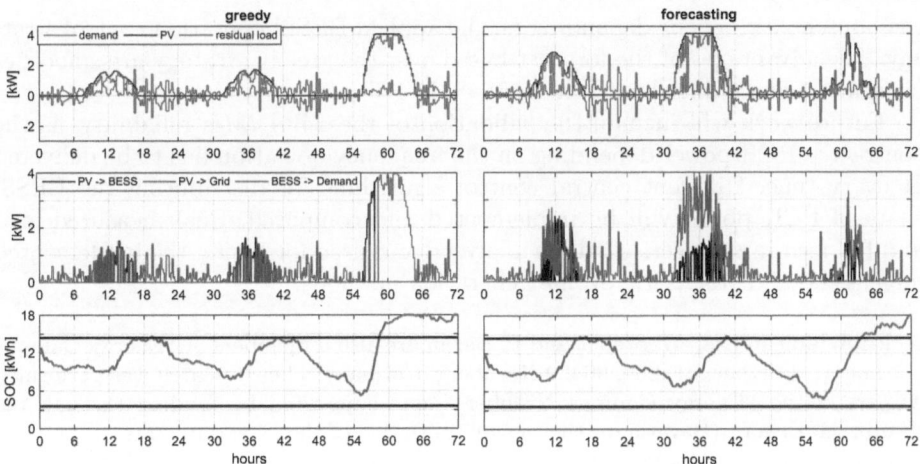

Fig. 5. PV, load and SOC time-series for greedy and forecast based control strategy

In conclusion, there is a significant benefit for the household by installing a BESS, but applying the forecasting based and grid supporting strategy instead of the greedy one causes no significant effect. The slight differences in Table 1 can be explained by the stochastic variation of the input data.

Figure 5 shows exemplary simulation results for the greedy (forecast based) algorithm on the left (right). The graphs of both strategies differ, as the PV and load profile generation is stochastic and leads to different profiles in each run. Applying the forecast based control strategy avoids high power gradients of PV feed-in and leads to a smoother BESS' charging. Especially day three of the greedy graphs compared to day two of the forecast graphs in Fig. 5 shows this. At times of high solar feed-in, the BESS charges with significantly lower power and the surplus PV power is fed into the grid. Furthermore, there are not so high gradients from charging to non-charging during times of PV provision anymore. In the greedy control scenario they result from the fact, that charging stops as soon as the SOC is more than 80 % of the capacity. The fluc tuations in both strategies come from providing PCR power according to the frequency in Fig. 2 and Eq. 1. This of course effects the power that can be stored in or taken from the BESS from/for the house. The truncated PV system power comes from EEG regulations, which bound the PV feed-in to 70 % of its peak power.

5   Conclusion and Outlook

This *Research in Progress Paper* presented a simulation approach for a real installed virtual mass storage composed of small BESS placed in households with a roof-top PV system for simultaneous internal consumption maximization and PCR power provision. The latter is coordinated by a central swarm controller. First simulation results show good working of implemented methods and a household's benefit of installing a BESS. There is no significant influence on the household's benefit depending on the applied BESS' operation control strategy. The advantages of the forecast based over the greedy strategy are smoother charging and avoiding high peaks of feed-in PV power.

Future work will include the validation of the *SWARM*'s reliability, as the contracted PCR power depending on the frequency deviation has to be delivered at every time. Different central control algorithms for distributing the BESS' share of PCR power will be implemented and compared. Real measured load and PV feed-in time series will be analyzed and used for fitting the implemented forecasting and stochastic profile generation methods.

Acknowledgments. D. Steber and P. Bazan are also a members of "Energy Campus Nürnberg", Fürther Str. 250, 90429 Nürnberg, Germany. Their research was performed as part of the "Energy Campus Nürnberg" and supported by funding through the "Aufbruch Bayern (Bavaria on the move)" initiative of the Bavarian state.

References

1. 50Hertz Transmission GmbH, Amprion GmbH, TransnetBW GmbH, TenneT TSO GmbH: Eckpunkte und Freiheitsgrade bei Erbringung von Primaerregelleistung. Technical report, April 2014. www.regelleistung.net
2. Bazan, P., Luchscheider, P., German, R.: Rapid modeling and simulation of hybrid energy networks. In: Proceedings 2015 SmartER Europe Conference Essen, Germany, February 2015
3. Beck, H.P., et al.: Studie Eignung von Speichertechnologien zum Erhalt der Systemsicherheit. final report, Energy Research Centre of Lower-Saxony (efzn), Goslar, Germany, March 2013
4. Borsche, T.S., Ulbig, A., Andersson, G.: Impact of frequency control reserve provision by storage systems on power system operation. In: Proceedings 19th IFAC World Congress, vol. 19, pp. 4038–4043. IFAC, Cape Town, August 2014
5. Deutsche Energie-Agentur GmbH (dena): dena ancillary services study 2030. requirements for a secure and reliable power supply with a high percentage of renewable energy. Technical report, dena, Berlin, Germany, July 2014
6. German Federal Ministry of Justice and Consumer Protection: Renewable Energies Act, July 2014. (Last changes on 29th June 2015. www.juris.de)
7. Hill, C., et al.: Battery energy storage for enabling integration of distributed solar power generation. IEEE Trans. Smart Grid **3**, 850–857 (2012)
8. Houwing, M., Negenborn, R., De Schutter, B.: Demand response with micro-CHP systems. Proc. IEEE **99**(1), 200–213 (2011)
9. Megel, O., Mathieu, J., Andersson, G.: Scheduling distributed energy storage units to provide multiple services. In: Power Systems Computation Conference (PSCC), pp. 1–7, Wroclaw, Poland, August 2014
10. Moshoevel, J., et al.: Analysis of the maximal possible grid relief from PV-peak-power impacts by using storage systems for increased self-consumption. Appl. Energy **137**, 567–575 (2015)
11. Oldewurtel, F., et al.: A framework for and assessment of demand response and energy storage in power systems. In: Proceedings of the Bulk Power System Dynamics and Control - IX Optimization, Security and Control of the Emerging Power Grid, IREP Symposium, pp. 1–24, August 2013

12. Weniger, J., Tjaden, T., Quaschning, V.: Sizing of residential PV battery systems. Energy Procedia **46**, 78–87 (2014). 8th International Renewable Energy Storage Conference and Exhibition (IRES 2013)
13. WIP Renewable Energies: Facilitating energy storage to allow high penetration of intermittent renewable energy. Intelligent Energy Europe Programme of the Europan Union, April 2015. www.store-project.eu

Hierarchical Simulation of the German Energy System and Houses with PV and Storage Systems

Peter Bazan[✉], Marco Pruckner, David Steber, and Reinhard German

Computer Networks and Communication Systems,
University of Erlangen-Nuremberg, Erlangen, Germany
{peter.bazan, marco.pruckner, david.steber,
reinhard.german}@fau.de

Abstract. The increase of renewable energies leads to new solutions in the field of decentralized energy storage. Houses with photovoltaic systems and battery storage systems can provide services for the power grid. But the isolated examination of only a few houses neglects the interaction of the houses with the power grid. We combine a model of the German energy system and a model of houses with photovoltaics and batteries. The two coupled hierarchical simulation models are then used to study different scenarios regarding the extension of renewable energy sources in Bavaria. Due to differences between the forecasted and real residual load and restrictions in the transmission grid, provision of control power is needed. The case studies show the amount of control power that will be provided by the houses with battery storage systems. In addition, the impacts on the electricity costs per year for a house are shown.

Keywords: Hierarchical modeling · Electrical energy system · Renewable energy · Control power · Storage system · Model aggregation · Hybrid simulation

1 Introduction

Energy system models are very important to support various stakeholders during the energy transition. These models can help to analyze and assess different configurations of the energy system in the future with respect to the share of electricity generated by renewable energy sources (RES), the extension of storage facilities or the development of thermal generation units. Hence, possible risks and miscalculations can be recognized at an early stage.

On the other hand, houses are increasingly being equipped with photovoltaic systems (PVs). Their fluctuating power generation has a great influence on the previously central power supply. To mitigate the problems, houses can be equipped with batteries which can be used to maximize the internal consumption of the generated renewable energy. But the current research also deals with how these batteries can be used to generate a benefit for the home owner and the power grid.

Here, however, the question arises if the use of these batteries is still possible if not only individual households are equipped with batteries, but a large number of

© Springer International Publishing Switzerland 2015
S. Gottwalt et al. (Eds.): EI 2015, LNCS 9424, pp. 12–23, 2015.
DOI: 10.1007/978-3-319-25876-8_2

households. With a surplus of PV energy in the households and the power grid, we probably can assume that the batteries are already fully charged and can't provide the necessary control power for the grid. To analyze such problems, we present a hierarchical simulation of the German energy system coupled with a large number of houses with PV and storage systems.

The simulation model of the German energy system consists of the most relevant parts of this system on a high level. Components for the electricity demand, the electricity generated by renewable energy sources (wind, photovoltaic, hydro power, biomass), thermal generation units, and storage facilities are integrated. The electricity demand, as well as the renewable energy feed-in, is based on a stochastic model. Due to restrictions in the transmission grid and differences between the forecasted and real residual load, provision of control power is needed and provided by some of the thermal generation plants and the storage facilities.

Due to the growing number of homes with photovoltaic and battery storage systems, the batteries can also be used for the provision of control power. The simulation model of hundreds of homes with photovoltaics and batteries models the electricity demand and PV power generation of each home and the corresponding state of charge of the battery. A surplus of energy - after calculating the internal balance of electricity generation and demand - is stored at the battery if possible, a shortage of energy is compensated by discharging the battery. A remaining imbalance is compensated by the power grid. After the coupling with the German energy system model, the houses also provide control power, reducing the amount of control power generated by some of the thermal generation plants and the storage facilities. The two coupled hierarchical simulation models are then used to study different scenarios regarding the extension of renewable energy sources in Bavaria. The case studies show the amount of control power that will be provided by the houses with battery storage systems. In addition, the impacts on the electricity costs per year for a house are shown.

2 Related Work

This paper presents the coupling of a micro-scale smart grid simulation approach for homes equipped with PVs and batteries with a macro-scale electrical system simulation of the German energy system. Established model coupling approaches focus on coupling different system areas. In [1] an economic energy market model is used in combination with an electricity grid simulation. There are considered restrictions for power plants, storages and RES as well as network constraints. Another approach is presented in [2], where real-time pricing is given as an example of combining economical and technical aspects within one simulation model. A tool for controlling the data-flow between different simulators and simulations of combined models and control strategies is the mosaik platform [3] for complex simulation models of energy systems and smart grids.

Another way of doing simulations of cross-domain specific problems is to embed detailed models (e.g. of batteries like in [4]) into given simulation frameworks. Furthermore, different simulation frameworks can be used in order to solve specific problems (e.g. energy management, grid simulation). For smart energy systems, the co-simulation of the energy system and the ICT layer can be carried out in order to improve and test control strategies, like it is done in [5]. There is presented the coupling of continuous simulation of power systems and discrete event based simulation of communication in combination with automation and control systems.

Basically, there are many publications in the field of modeling energy systems, which are relevant to the subject of this paper. In Connolly et al. [6] an overview of 37 energy system models is given. The models can be used to analyze the integration of electricity generated by renewable energy sources. The energy models can be differentiated by their applied methodology (simulation, optimization, etc.), geographical area (local, national, worldwide), scenario timeframe (one year, more years, no limit), and time-step (minutes, hourly, yearly). The authors came to the conclusion that there is no energy model available which addresses all issues related to integrating renewable energy. Generally, the applied methodology, geographical area, scenario timeframe and time-step depend on the issue which will be answered with the energy system model (see also [7]). For example, if we are interested in determining an optimal capacity expansion plan for renewable energy sources with respect to economic constraints, than we usually use an optimization framework; if we are interested in comparing different framework conditions and their impacts on the energy system, than we use a simulation-based energy model. In [8] for the strategic energy system planning a hybrid modeling concept is derived. The short-term fluctuations are analyzed in a simulation model, whereas the long-term development of the energy system is assessed in an optimization model. The developed simulation model has a high temporal resolution on shorter time scale and can be used to simulate the power plant dispatch over a maximum of one year on a national geographical area.

3   Simulation Frameworks and Coupling

In this section we describe the two used simulation frameworks and their coupling. The first simulation framework models the German energy system. The second framework is for the construction of smart energy grids.

3.1   Simulation Framework for the German Energy System

The simulation framework for the German electrical energy system is a comprehensive energy system model with a focus on the federal state Bavaria. Components for the electricity demand, the electricity generated by renewable energy sources (wind, photovoltaic, hydro power, biomass), thermal generation units, and storage facilities are integrated. The modeling of these components is described in various publications [9–11]. For instance, in [10] the modeling of the electricity demand is explained. For the demand modeling it contains a stochastic model based on the published load profiles from the ENTSO-E (European Network of Transmission System Operators for Electricity). In [11] the modeling of the feed-in of highly fluctuating photovoltaic systems and wind energy plants can be found. For the modeling of the feed-in of photovoltaic and wind energy plants a stochastic model based on official data provided by the German transmission system operators 50Hz, Amprion, TenneT, and Trans-netBW is included.

A basic overview of the most relevant input and output parameters for the simulation framework is depicted in Fig. 1. Apart from the demand profiles and the feed-in structure of different renewable energy sources, a comprehensive set of input parameters to control the behavior of different components and the interaction between them is available.

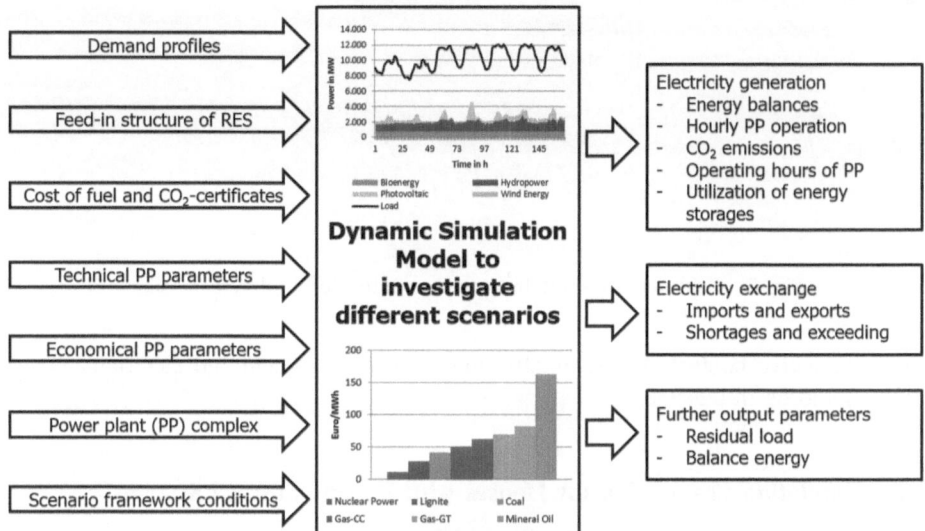

Fig. 1. Input and output parameters for the simulation framework

On the right side of Fig. 1 the output parameters are shown. In order to investigate the balance energy, forecasts for the feed-in of wind energy plants and photovoltaic systems as well as for the electricity demand are implemented. The framework is based on the commercial simulation tool AnyLogic 7 [12], which is written in JAVA. The investigated geographical area is limited to Germany and Austria.

In order to consider transmission grid restrictions between the different regions of Germany, a multipoint modelling approach is used. The geographical area and the division in different points are depicted in Fig. 2. Due to differences between the forecasted and real residual load, the provision of control power is also taken into account. Currently, control power can be provided by electricity storage facilities and thermal generation units. If the model is coupled with the simulation framework for houses, the control power is at first provided by the houses' batteries. If more electricity is generated than needed, the surplus is stored in the houses' batteries, the electricity storage facilities, or the generation of thermal generation is down-regulated. Conversely, if the real residual load is higher than the forecasted one, positive control power is needed. In this case first the houses' batteries are discharged, then the electricity storage facilities are discharged, or the thermal generation units can be started to

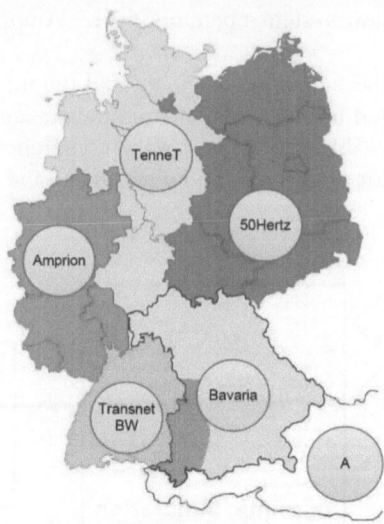

Fig. 2. Considered geographical area and division in different points

provide positive control power. Finally, the electricity demand and electricity generation should be in balance.

3.2   Simulation Framework for Houses with PV and Storage Systems

The simulation framework i7-AnyEnergy [13] is designed for the fast construction of smart grid energy system models with renewable energy sources and storage systems [14]. It is based upon the hybrid simulation tool AnyLogic 7 [12] and utilizes its state chart paradigm for control decisions and its system dynamics paradigm for energy and cost flows. From its build in components for energy demand, PV, battery, controller, and costs a house can be constructed (Fig. 3, left) using the interface/filter concept described in [13] for the coupling of components.

The strategy for the internal house controller without a control power request from the German energy system simulation is to first use the energy from the PV for the internal demand. A remaining surplus charges the battery; a remaining demand discharges the internal battery. If the battery can't be charged or discharged because of its state of charge or restrictions in the charging/discharging power, the remaining electricity is bought from or sold to the power grid. This strategy maximizes the internal consumption of the energy from the PV.

On the other hand, with a request of control power from the German energy system simulation at first the internal balance of the energy from the PV, the internal demand, and the requested control power is computed. Again a remaining surplus charges the battery; a remaining demand discharges the internal battery. If the remaining charging/discharging power of the battery is less than required, it is at first used for the residual power of the internal demand and PV balance. Only from this balance

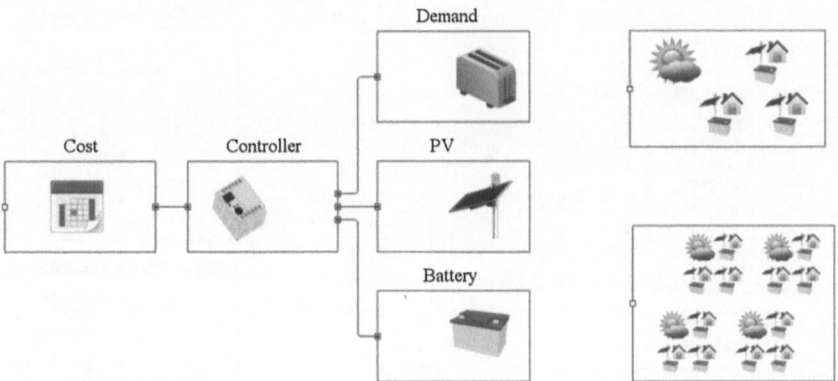

Fig. 3. Components of the simulation framework i7-AnyEnergy: a house with PV and battery (left), local region with one hundred houses with one weather model for the solar irradiation (right top), four local regions (right bottom)

remaining electricity is bought from or sold to the power grid. The control power that can't be used or charged/discharged internally can't be bought or sold, of course.

A group of one hundred houses is connected to a weather model for the solar irradiation (Fig. 3, right top). For the simulation of four different solar irradiations for houses in four different regions, the four groups of houses are equipped with different weather models (Fig. 3, right bottom).

3.3 Model Coupling

The simulator i7-AnyEnergy and the simulation framework for the German electrical energy system are coupled in order to investigate the provision control power by charging or discharging the houses' batteries. The two simulation models are implemented with AnyLogic and can therefore be combined as two components of a superordinate simulation model (Fig. 4) using the interface/filter concept described in [13].

The component *Write* is responsible for writing the results to disk and the component *Graph* for displaying intermediate results during the runtime. The component *Experiment* contains the parameters for different simulation runs, e.g., different battery capacities or PV peak powers. The System Dynamics component *Controller* interconnects and synchronizes the two models *Germany energy system* and *Houses with PV and battery*. The *Controller* scales the simulation values of the 400 houses up to 200,000 houses and connects them to the model of the federal state of Bavaria. The 200,000 houses correspond to 25 % of about 800,000 households with four or more persons in Bavaria.

The details of the component *Controller* are shown in Fig. 5. The control power surplus of 20,982 kW in the Bavarian part of the German energy system (*e_net_balance*, calculated by *Germany energy system*) is scaled down to the 400 houses (*e_net_houses_bal*) of the house simulation model. The *Controller* askes the houses to

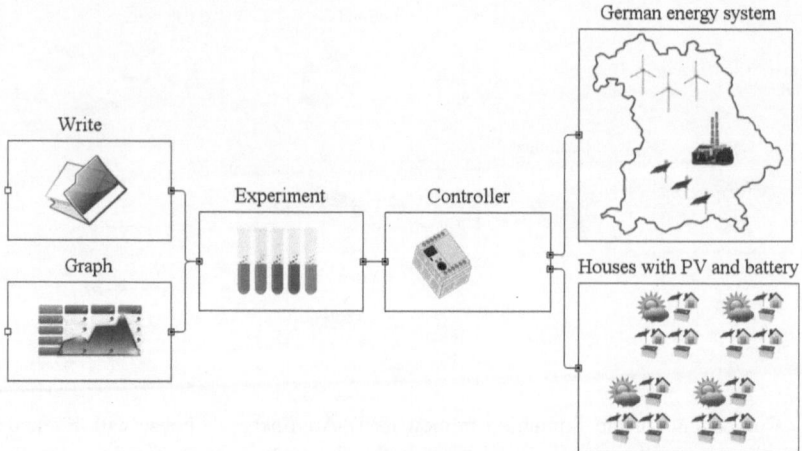

Fig. 4. Components of the hierarchical simulation model of renewable energy systems with storage

charge their combined batteries with 41.964 kW (*e_net_houses_r*) – the maximal combined charging power at the moment is 1600 kW – and the function *calc_e_ net_houses_r* distributes the power among the houses according to their reported individual remaining charging/discharging power. The houses agree to charge their batteries with the requested power of 41.964 kW (*e_net_houses_a*, calculated by

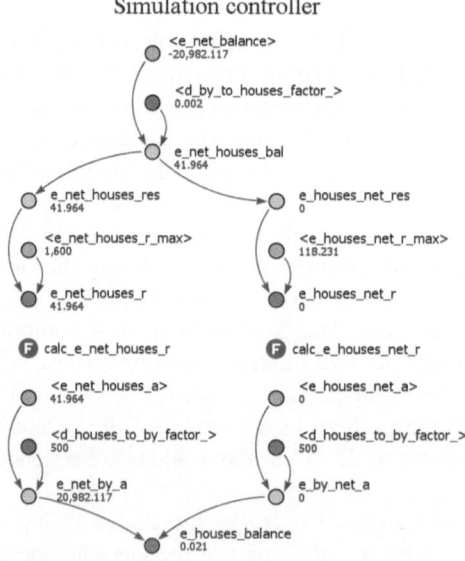

Fig. 5. The System Dynamics model of the component *Controller* coupling the German energy system simulation and the simulation of the houses with PVs and batteries

Houses with PV and battery). This value is scaled up to 200,000 houses (20,982 kW for *e_net_by_a*) and converted to GW (*e_houses_balance*) for the *Germany energy system* component.

For this snapshot the whole control power surplus of the energy system can be compensated by the houses; no control power has to be provided by electricity storage facilities or thermal generation units of the component *Germany energy system*.

4 Hierarchical Simulation of the Coupled Models

The described hierarchical simulation model is used for the analyses of different scenarios. After the definition of the parameters, simulation results for the energy costs for a house and the fractions of delivered control power with respect to the German energy model's requests of control power are given.

4.1 Basic Assumptions

In order to investigate the impact of many battery storages in households on a larger scale, we have to define some basic assumptions for the year 2023. The year 2023 is the first year in Germany without nuclear energy. The framework conditions for Germany without Bavaria can be found in [15]. The scenario framework of the German Netzentwicklungsplan [15] also provides a list of thermal generation units which are in service in 2023. For Bavaria itself we define three different configurations of the installed power for wind energy plants, photovoltaic systems, biomass-fired plants, hydro power plants and geothermal power plants which are shown in Table 1.

Table 1. Different configurations for the extension of renewable energy sources in Bavaria

	Total installed power in the year 2023 (in MW)				
	Hydro power	Photovoltaic systems	Wind power plants	Biomass-fired plants	Geothermal power plants
2013 (ref.)	2,900	10,562	1,125	1,171	22
Pessimistic	2,900	13,444	2,521	1,245	300
Realistic	2,900	15,489	5,029	1,456	300
Optimistic	2,900	22,105	6,559	1,838	300

The pessimistic and the optimistic scenarios are used for the simulation runs and each scenario is combined with different scenarios for the households (Table 2). The row *Reserved* gives the state of charge of the battery in percent that is reserved for the power grid. A house can't use the energy stored in the battery for its own demand if the state of charge is lower than the reserved state of charge. The costs of energy from the power grid is 0.29 €/kWh. The PV feed in tariff for PV energy exported to the grid is 0.125 €/kWh. The costs regarding to energy charged or discharged because of balancing power are set to 0 €/kWh. Therefore the energy costs of a house after one

Table 2. Configurations for the maximal charging and discharging power, the capacity of the battery, the reserved capacity for the purpose of power compensation, and the peak power of the PV of a house

Battery	Dis-/Charge [kW]	8				12			
	Capacity [kWh]	10				20			
	Reserved [%]	0	25	50	75	0	25	50	75
PV	Peak power [kW]	3				6			

year with respect to the costs of a house that doesn't provide control power reflect the costs of providing control power.

4.2   Simulation Results

The hierarchical model was simulated with the previous described 24 different parameter sets over a period of one year. The electricity energy costs for one house are given in Fig. 6. If the house is equipped with a 10 kWh battery, a 3 kW PV, and no

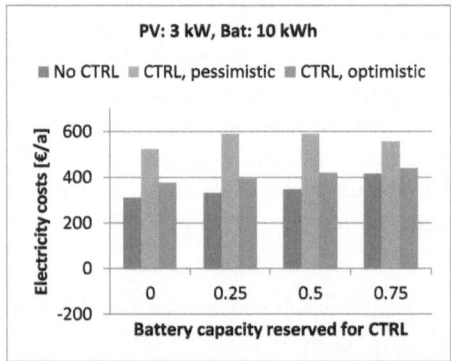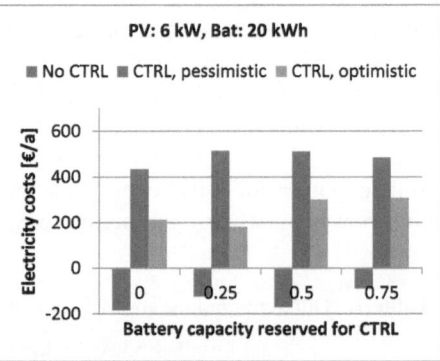

Fig. 6. Mean electricity costs per year for a house with respect to different scenarios

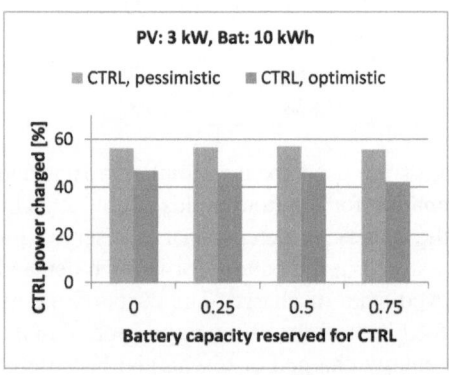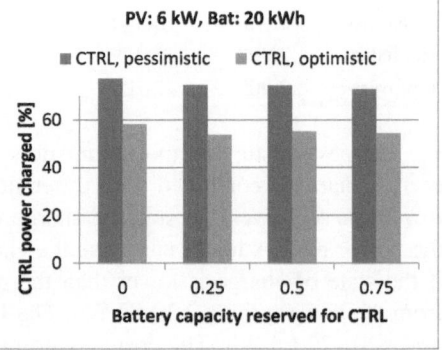

Fig. 7. Fraction of the requested negative control power charged into the batteries

battery capacity is reserved, the costs increase by approximately 200 €/a in the case of the pessimistic scenario and by approximately 70 €/a in the case of the optimistic scenario with respect to the case where no control power is provided. If the house is

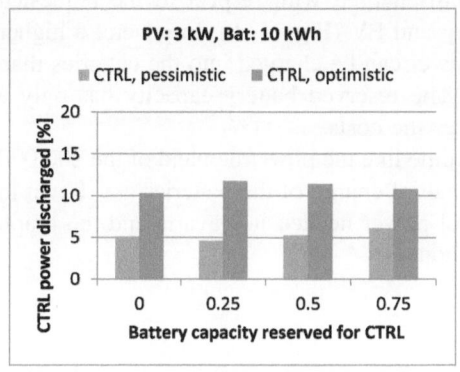
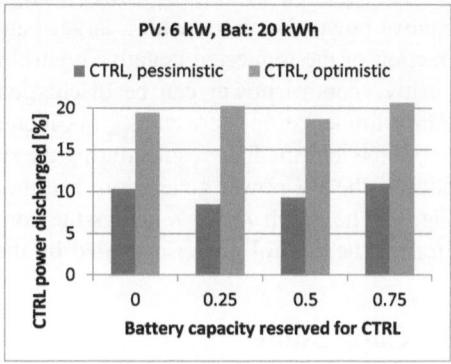

Fig. 8. Fraction of the requested positive control power discharged from the batteries

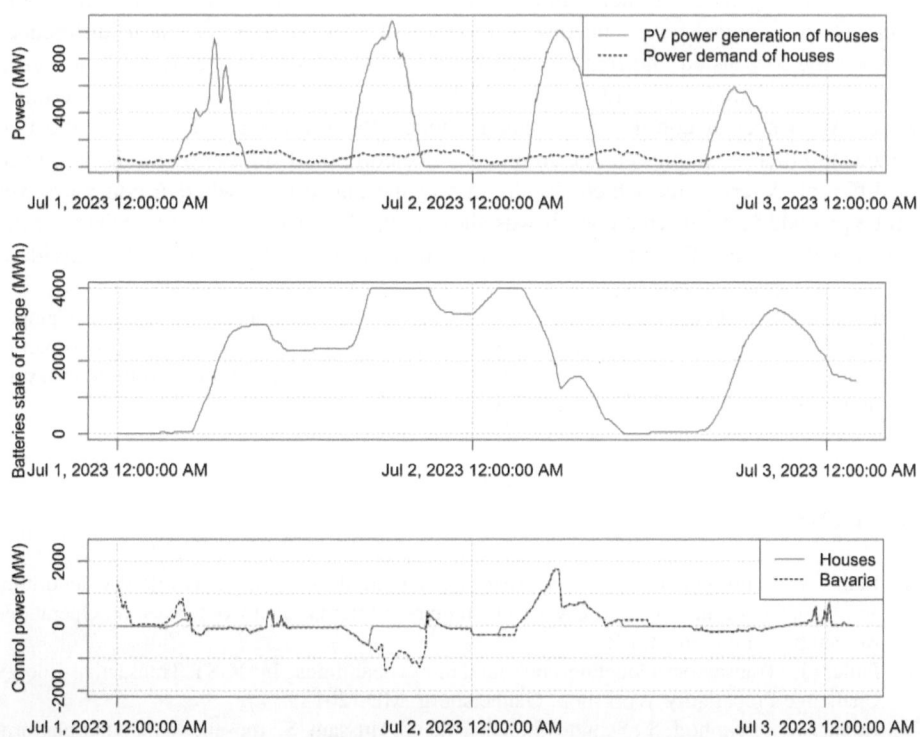

Fig. 9. The Graph component containing plots of the power demand and PV power generation of 200,000 houses, their combined state of charge of the batteries, and the needed control power of Bavaria and provided control power of the houses

equipped with a 20 kWh battery, a 6 kW PV, and no battery capacity is reserved, the costs increase by approximately 620 €/a in the case of the pessimistic scenario and by approximately 400 €/a in the case of the optimistic scenario with respect to the case where no control power is provided.

The fraction of control power charged/discharged with respect to the requested control power increases with a larger battery and PV (Figs. 7, 8). In general a higher fraction of the requested negative control power can be charged into the batteries than positive control power can be discharged. The reserved battery capacity has only a minor influence on these values but increases the costs.

Some intermediate results during the runtime like the power demand of the 200,000 houses, the PV power generation, and the state of charge of the batteries are shown in Fig. 9. The graph *Bavaria* shows the control power needed in Bavaria and the graph *Houses* the control power provided by the houses.

5   Conclusion

The described simulation model of the German energy system and the simulation model of 400 houses with PV and batteries scaled up to 200,000 houses have been coupled forming a hierarchical simulation model. The German energy system model calculates the control power needed in the federal state of Bavaria due to differences between the forecasted and real residual load and restrictions in the transmission grid. The batteries of the houses are used to fulfill the demand of positive or negative control power. With such a coupled hierarchical model, the mutual interference of the two models and their strategies can be simulated. The combined simulation model was used for different experiments, which should answer the question of whether control power can be provided and at what cost. It was shown that the costs depend on the size of the battery and PV, and that more negative than positive control power can be provided.

Acknowledgment. Peter Bazan and David Steber are also members of "Energie Campus Nürnberg", Fürther Str. 250, 90429 Nürnberg, Germany. Their research was performed as part of the "Energie Campus Nürnberg" and supported by funding through the "Aufbruch Bayern (Bavaria on the move)" initiative of the Bavarian state.

References

1. Raths, S.: Multi level european electricity market simulation using network flow algorithm and lagrangian relaxation. In: OR2013 Conference: International Conference on Operations Research, Rotterdam (2013)
2. Fuller, J.: Transactive modeling and simulation capabilities. In: NIST Transactive Energy Challenge Preparatory Workshop, Gaithersburg, MD (2015)
3. Rohjans, S., Lehnhoff, S., Schütte, S., Scherfke, S., Hussain, S.: mosaik - a modular platform for the evaluation of agent-based smart grid control. In: 4th IEEE PES Innovative Smart Grid Technologies Europe (ISGT Europe), Copenhagen (2013)

4. Palensky, P., Widl, E., Stifter, M., Elsheikh, A.: Modeling intelligent energy systems: co-simulation platform for validating flexible-demand EV charging management. IEEE Trans. Smart Grid **4**(4), 1939–1947 (2013)
5. Stifter, M., Kazmi, J.H., Andrén, F., Strasser, T.: Co-simulation of power systems, communication and controls. In: 2014 Workshop on Modeling and Simulation of Cyber-Physical Energy Systems (MSCPES), pp. 1–6, Berlin (2014)
6. Connolly, D., Lund, H., Mathiesen, B.V., Leahy, M.: A review of computer tools for analyzing the integration of renewable energy into various energy systems. Appl. Energy **87**(4), 1059–1082 (2010)
7. Möst, D., Fichtner, W.: Einführung zur Energiesystemanalyse. In: Möst, D., Fichtner, W., Grundwald, A.: Energiesystemanalyse, pp. 11–31. Universitäts-Verlag Karlsruhe (2009)
8. Rosen, J.: The future role of renewable energy sources in European electricity supply: A model-based analysis for the EU-15, Universitäts-Verlag Karlsruhe (2008)
9. Pruckner, M., German, R.: A hybrid simulation model for large-scaled electricity generation systems. In: Pasupathy, R., Kim, S.H., Tolk, A., Hill, R., Kuhl, M.E.: Proceedings of the 2013 Winter Simulation Conference, pp. 1881–1892. Washington, D.C. (2013)
10. Pruckner, M., Eckhoff, D., German, R.: Modeling country-scale electricity demand profiles. In: Tolk, A., Diallo, S.Y., Ryzhov, I.O., Yilmaz, L., Buckley, S., Miller, J.A.: Proceedings of the 2014 Winter Simulation Conference, pp. 1084–1095. Savannah, GA (2014)
11. Pruckner, M., German, R.: Modeling and simulation of electricity by renewable energy sources for complex energy systems. In: Proceedings of the 47th Annual Simulation Symposium (ANSS 2014), pp. 1–9, Tampa, FL (2014)
12. AnyLogic 7. http://www.anylogic.com
13. Bazan, P., German, R.: Hybrid simulation framework for renewable energy generation and storage grids. In: Proceedings of the International Workshop on Demand Modeling and Quantitative Analysis of Future Generation Energy Networks and Energy Efficient Systems, Bamberg (2014)
14. Bazan, P, Luchscheider, P., German, R.: Rapid modeling and simulation of hybrid energy networks. In: Proceedings of the 2015 SmartER Europe Conference, Essen (2015)
15. 50Hertz Transmission GmbH, Amprion GmbH, TenneT TSO GmbH, TransnetBW GmbH: Szenariorahmen für die Netzentwicklungspläne 2015 – Entwurf der Übertragungsnetz betreiber (2014)

Energy Service Description for Capabilities of Distributed Energy Resources

Tim Dethlefs[1]([✉]), Christoph Brunner[2], Thomas Preisler[1], Oliver Renke[1], Wolfgang Renz[1], and Andrea Schröder[3]

[1] Faculty of Electrical Engineering, HAW Hamburg, 20099 Hamburg, Germany
tim.dethlefs@haw-hamburg.de
[2] It4power, 6300 Zug, Switzerland
[3] Forschungsgesellschaft Für Elektrische Anlagen und Stromwirtschaft E.V. (FGH), 68219 Mannheim, Germany

Abstract. The increasing number of volatile Distributed Energy Resources (DERs) in the electricity grid implies a rising level of complexity and dynamics. The integration and management of these DERs have lead to the introduction of the aggregator role, with the aim of providing energy services to system operators and the market. With regard to the often changing capabilities of DERs, the dynamical aggregation of DERs to meet the demand is still a matter of concern. In this paper a generic description for the capabilities of DERs will be introduced in order to allow the aggregator to efficiently search and find DERs suitable for aggregation. These reduced as possible and abstracted descriptions of the DER capabilities are called Energy Services, which should be complete enough for the aggregators search demands.

The Energy Service definition will be part of a recent research project, the Open System for Energy Services (OS4ES) that is going to enable the aggregator to control dynamically configured large scale Virtual Power Plants with IEC 61850. The results of this project and its field test should contribute to the further development of IEC 61850.

1 Introduction

The rising number of heterogeneous and volatile Distributed Energy Resources (DER) in the electricity grid leads to an increasing complexity in grid management [2]. The aggregator role, discussed in several research studies and standards and currently establishing in the energy domain, shall serve as a complexity managing entity [5]. It combines DERs of different characteristics to so called Virtual Power Plants (VPP) for the participation in larger energy markets [7]. The business processes between aggregator and Smart Grid Actors (e.g. system operators and Balance Responsible Parties) are currently under development in many projects (e.g. USEF-project [3]). Still a matter of concern is the interaction and information exchange between aggregators and the DER systems. Classical aggregator concepts assume an often static set of DER systems which the aggregator can use to provide services for the participants. With the rising

© Springer International Publishing Switzerland 2015
S. Gottwalt et al. (Eds.): EI 2015, LNCS 9424, pp. 24–35, 2015.
DOI: 10.1007/978-3-319-25876-8_3

number and variety of DERs, business models, and often changing power flows we assume that in the future aggregators must be able to dynamically reconfigure their portfolios to serve the needs of the electricity domain and to react on flexibility requests from, e.g. the system operator. Furthermore, we assume that in future smart markets, more than one aggregator needs to be given access to available capabilities of a DER system for optimal use of the available resources. For this purpose, the aggregator must be able to find DER systems that provide the needed capabilities as Energy Services in a Smart Market [4]. For such Smart Markets, an active Registry System for Energy Services as described in [6] is necessary that allows the DER system providers to offer the capabilities of their DER systems to aggregators that can search and find these to aggregate them and thus, to participate in larger energy markets (e.g. energy exchanges, ancillary services, or Over The Counter [OTC] trades), as stated in Fig. 1.

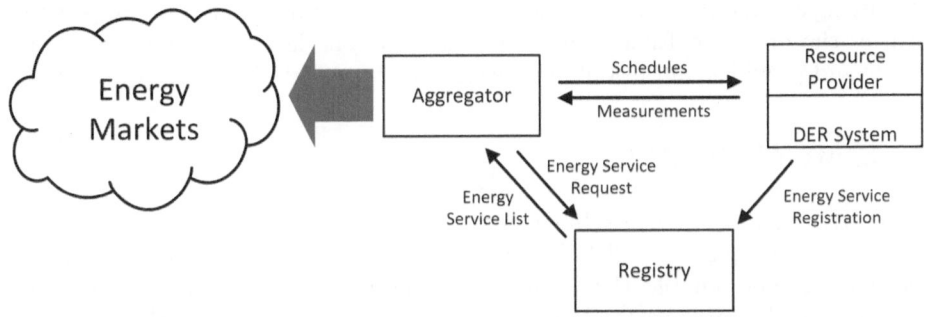

Fig. 1. Interaction between aggregators and DER systems using a common registry system for the exchange of dynamic energy service data. The aggregator utilizes the DER systems based on the energy service descriptions to participate in larger energy markets.

In order to offer often fluctuating DER capabilities that may change due to, e.g. environmental influences to aggregators, a common and generic description of these capabilities is required. This description model should contain as minimum information as possible for several reasons, e.g. data storage requirements and privacy and security concerns. Finally, the aggregator should handle the complexity of the heterogeneous DER systems by searching for relatively generic characteristics of the provided Energy Services. It has to be investigated how far the characteristics of a DER system can be abstracted from the technical base without leaving out necessary information for the aggregator. In this case it is important to distinguish between the search for Energy Services, where the informational complexity should be reduced to provide an easy and efficient search mechanism for the aggregator and the operation of a DER system. In the operational phase, it is likely that the complexity of the DERs cannot be made fully transparent. Recent standards like IEC 61850-7-420 work on semantic

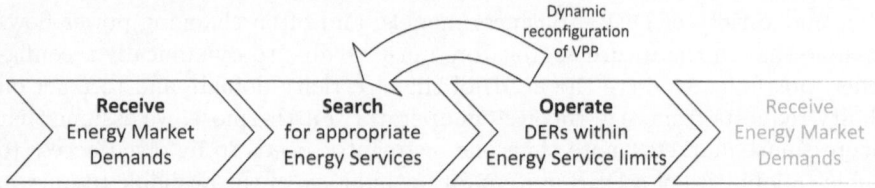

Fig. 2. Phases of the aggregation of a virtual power plant.

data models for DER systems to handle this challenge [1]. The Energy Service definition in this paper will focus on the search phase (see Fig. 2).

The remainder of the paper is structured as follows: in the next Section a state of the art overview on Energy Markets and established interaction models is given. In Sect. 3 the energy Service will be described and defined in the context of existing ancillary services. Section 4 focuses on an implementation project that utilizes the proposed Energy Service definition. The last Section concludes the paper and provides an outlook on recent research efforts and open questions.

2   State of the Art

Currently, existing energy exchange market systems, namely the European Power Exchange Spot market (EPEX SPOT SE) or the German internet platform regelleistung.net, enable the exchange of energy services and the settlement of commercial processes for larger market parties. Members of these auction-based markets, balance responsible parties, system operators and large resource providers (producers, consumers and prosumers), are able to participate in the day-ahead or the intraday markets where energy-orders can be placed and contracts concluded.

Independent from these auction based markets, the market parties may conclude OTC-transactions. Focusing on established energy market player and high-volume producer these market models do not provide easy access for small and medium-sized DERs besides the emerging VPP-concept.

Trying to fill this gap, governmental and non-governmental organizations are developing and publishing approaches which tackle the growing dissemination of DERs and as a consequence thereof the growing number of DERs willing to gain access to the energy markets. One example is the Universal Smart Energy Framework (USEF) [3], which focusses on an open framework containing specification, design and implementation guidelines so that participants or stakeholders are able to create a fully functional smart energy system. Within the projects scope, aggregators can accumulate flexibility and active demand and supply of DERs as well as of smaller aggregators to maximize the value of their energy generation and vend it in the integrated market. Energy Service Companies (ESCo) provide auxiliary services to the Prosumers, e.g. insight services or energy management services. Providing this features USEF may integrate

DERs in the integrated markets but still misses functionalities for aggregators to accumulate DERs directly.

Aligning with the governmental Data Access Manager (DAM)-concept [12] the Energy Service definition proposed here may fill this gap by providing this missing functionality for aggregating DERs by allowing a search by aggregators. The OS4ES-Project (see also Sect. 4) defines generic communication and Energy Service models while using a distributed registry system for storing this information. By enabling the resource povider to register and advertise his DER respectively its Energy Services conveniently as well as providing a centralized Service-search and -booking functionality for an aggregator to adjust or reconfigure his portfolio according to current demands, the OS4ES creates a dynamic system to utilize, accumulate and trade the capabilities of these low-volume producers in the manner of a VPP.

3 Energy Services

This Section should describe the scope of the Energy Services and propose a definition of their characteristics.

3.1 Scope

The proposed Energy Service definition shall cover the the provision of power that needs communication between the aggregator role and the resource provider. The aggregator should use a common registry system such as proposed in [6] to find Energy Services provided by DERs. The aggregator is then able to provide services for other actors, such as system operators, or balance responsible parties (BRPs) as for example described in the USEF-project.

For the aggregator role, three main use-cases are currently identified:

1. **Scheduling of power:** the most common use case, to provide active or reactive power within the context of a VPP.
2. **Ancillary Services:** there are several ancillary services of interest for the aggregator role, especially
   (a) Frequency Control: the aggregator provides a VPP that can perform Frequency Control. For this, DERs are needed that have the capability to measure the frequency and can react on deviations.
   (b) Volt/Var optimization: the aggregator provides the DSO a VPP capable of optimizing the Voltage. Also for this, DERs require special abilities.

In Sect. 3.2 it will be shown that the description of power in conjunction with additional information (e.g. energy profiles and forecasted data) is sufficient for most of the described use-cases within the context of the recent grid management.

The Energy Service definition must be distinguished from other service definitions i.e. the Service-terminology in computer sciences [9]. Energy Services are the representation of physical capabilities that are in general limited and thus not repeatable. Therefore, an Energy Service is a consumable object, which means when an aggregator reserves the Energy Service by expressing the intention to use it and then activates the capability (i.e. uses it), the Energy Service is no longer available for other parties.

3.2   Definition

For the classification of the proposed Energy Service definition it is important to determine the taxonomy and relation of the definition within the Smart Grid terminology. As stated in Fig. 3A, the central role in this context is the DER system. Every DER system is connected to the grid through at least one Point of Common Coupling (PCC) or grid connection point, so each PCC has a set A of logical DER systems. Every PCC is connected to a zone, i.e. the grid-zone of the according system operator. In Europe, this can either be the transmission grid or the distribution grid.

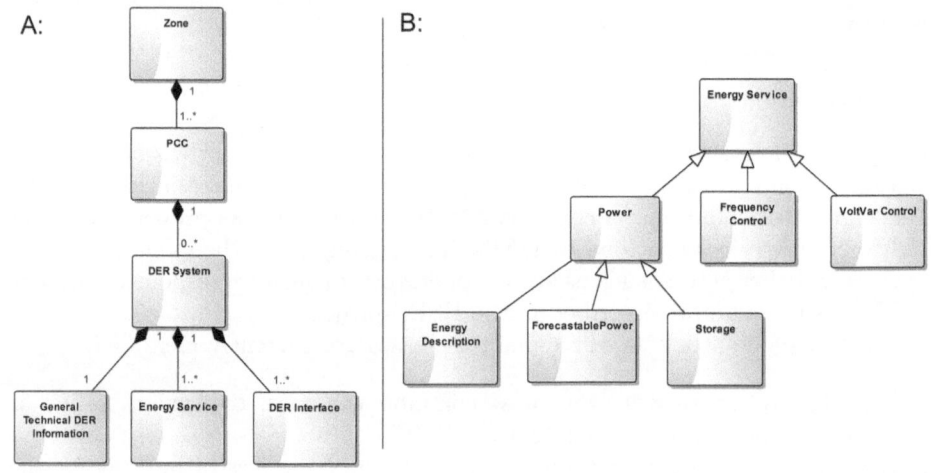

Fig. 3. A: UML-class diagram of the relation between the energy service and the organizational entities. **B:** Example structure of an energy service

A DER system $a_i \in A$ represents a physical or logical device and is composed of three different datasets. The general technical DER information contains all data relevant for the DER system, e.g. PCC and position data, owner, type of installation etc. The interfaces provide information how to communicate with a DER system and how to control it. Each DER system that can provide services for the grid holds a set of Energy Services S_i. The Energy Services $s_i \in S_i$ can be reserved and consumed by the aggregator. It is possible that a DER provides multiple Energy Services at the same time which may exclude each other. The set of rules C_i, describing how the parallel Energy Services influence each other must be defined by the resource provider and stored at a Registry System for Energy Services, where they can be executed when an aggregator reserves an Energy Service. The execution of the rules manages the visibility of other Energy Services provided by the same DER system for search requests by other aggregators.

Based on the scope of the Energy Service definition described in the previous Section, information about the provided power (active and reactive) is considered as the most important aspect for most use cases. The description of the maximum power available over time may be sufficient to describe nearly unconstrained power plants such as gas generators for the search. Depending on the type of DER system and its characteristics, additional data could be necessary. A forecast-based DER such as a wind turbine may also provide forecasts on the expected power in wost-case/best-case corridors until a certain forecast border (see Fig. 4A). Such a DER can provide three informational horizons:

1. The actual power: P_{actual} measured at the DER system, could be used by aggregators to project the capability of the DER system
2. The forecast-based area: a DER system can provide information on the projected corridor of power available within the Energy Service. Although, for environmental-dependent DERs some risk could be included. Such forecast-models can just provide reliable information within a certain period of time.
3. Nominal-value-based: beyond the forecast border, no further information on the characteristics can be given. aggregators can just rely on the maximum nominal-value of power P_{max} a resource provider has defined for this specific Energy Service entity. This implies that the aggregator must consider the type and characteristics of the DER system as well as environmental conditions. Thus, the aggregator must be also able to access the general technical information of the DER system to get information on the DER type.

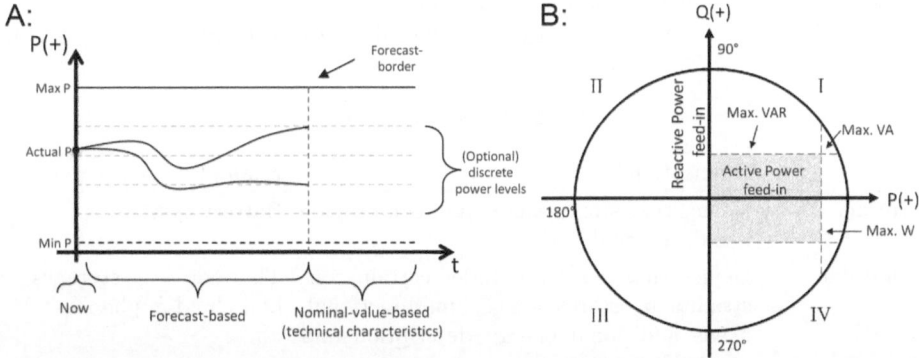

Fig. 4. A: Graphical description of a forecast-based energy service providing active power. **B:** Consumer-Producer reference frame reflecting active and reactive power relation, abridged from [11]

The proposed power-description model is applicable for both, active and reactive power, so that in some cases rules are needed, when a resource provider offers an Energy Service for active and reactive power in parallel. In the most simple case both Energy Services are totally decoupled, so the DER system can always offer an Energy Service with P_{max} and one with Q_{max}. Thus, the resulting complex power pointer S can be anywhere in the gray marked area of Fig. 4B. Defined as a rule, the resource provider can limit this area to reflect the technical abilities of the DER system, e.g. $|S| \ll \sqrt{P_{max}^2 + Q_{max}^2}$.

An aggregator can claim an Energy Service instance for a certain time-period, so the actor is allowed to send control signals within the limits of the reserved Energy Service i.e. $P_{min} \leq P_{demand} \leq P_{max}$ (for reactive power respectively). As the DER can be dependent on environmental conditions, the delivered power can be $P_{actual} \leq P_{max}$, which must be taken into account by the aggregator. In order to allow an optimal utilization of the DER system, the resource provider must be able to quantize the Energy Services, e.g. split the total nominal power of the DER system into multiple Energy Service instances.

Following the DER system definition in IEC 61850-7-420 it is assumed that a DER system is composed out of one or more DER Units and serves as an overlay and managing instance for all subsidiary DER Units. DER systems can be ordered hierarchically i.e. a DER system can be part of another DER system as a DER Unit. As the complexity of such systems may rise with the hierarchical depth of the system, it is recommended to logically decompose complex DER systems when describing the capabilities with Energy Services. Thus, the above described data will be more clearly defined, resulting in an easier search and the aggregator would be able to interpret the characteristics of the DER system better.

Depending on the class of the DER system in conjunction with the power profiles (see Fig. 4A), it is obviously that additional information are necessary to plan the devices. These information specialize the Energy Service description for the characteristics of the DER system while adding only little complexity to the search. In contrast to more market oriented DER-classifications, e.g. the FPAI-class model [8] the here proposed DER-class-model for Energy Services stated in Table 1 should strongly focus on the generalization of technical aspects. The

Table 1. Device classes

Type	Description	Example
Storage	Device that stores energy and can provide it bidirectional	Battery system
Shiftable	Device that can shift under certain constraints or stores energy unidirectional, needs additional energy description and maybe forecast model	Thermal storages, household appliances
Deterministic	Controllable device, can depend on forecasts-model	Gas generators, PV inverters

three classes of DERs are currently under consideration are storages, shiftable DERs and deterministic DERs.

Deterministic DERs can be almost fully described by their provided power and have only few additional constraints that can be expressed through further information within the power profile (e.g. the already described predictability with forecasting horizons).

The shiftable DER class includes shiftable consumer devices (e.g. household appliances) or thermal storages. For these are often forecast data necessary but also information on the total expected energy demand, the energy storage capacity, and the shifting characteristics. In this case, an additional dataset can be given as stated in Fig. 5. It describes the interrelation between the minimum power needed and the maximum power over time, determining a corridor of allowed energy consumption. Figure 5 can be a description of the daily energy demand of a household that has a minimum energy demand for maintaining the lowest temperature but also has the ability to consume additional energy.

For the storage class, covering classical battery systems with bidirectional power flows energy is also an important information on the capacity of the storage.

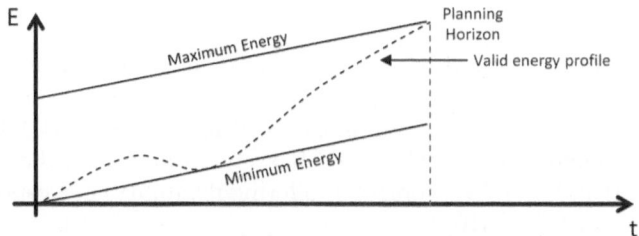

Fig. 5. Description of valid energy profiles, supporting the power description.

This class model allows the definition of an Energy Services class system (see Fig. 3B). The power-subclass describes deterministic DERs which can be specialized into forecastable DERs. The Energy Description as in Fig. 5 can be added when necessary. The Energy Service should not only realize the direct exchange of power and energy, but also on further use and business cases. The Energy Service *Frequency Control* uses the power description model together with additional information on the control energy capabilities (e.g. primary control capability, secondary and tertiary capabilities) while *Volt/Var* is comparable. Thus a large number of possible use cases for DER systems can be covered.

4   Case Study: The Open System for Energy Services

The EU funded research project Open System for Energy Services (OS4ES)[1] will provide a platform that allows aggregators to search and find matching

---

[1] http://www.OS4ES.eu.

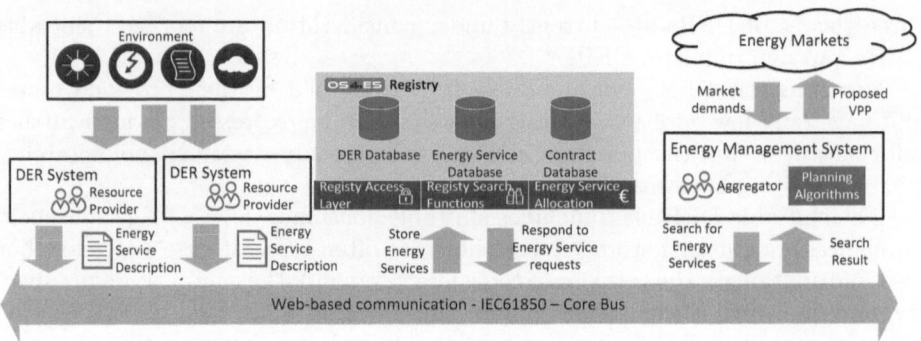

Fig. 6. OS4ES system

DER systems for energy service requests received from a DSO or TSO. Figure 6 shows how this will be achieved.

The resource provider of a DER system (e.g. a PV plant, a wind power plant, a combined heat and power plant, an electric vehicle or any combination of these) registers the data of its DER system in the OS4ES Registry. This DER system data can be categorized in:

- Data as it is can be found on technical data sheets (e.g. owner of the DER system, location, voltage level, point of common coupling, nominal power, type of DER system) corresponding to the White Pages of UDDI and
- Data for the energy service(s) the DER system offers (e.g. data relevant for primary control, Volt/VAR control or offering flexibility) corresponding to the Yellow Pages of UDDI [10].

According to these two categories the technical data sheet data is stored in the DER Database of the OS4ES Registry while the energy service related data is stored in the Energy Services Database. An aggregator receiving requests from the Energy Markets (e.g. flexibility request of 30 MW for the next day) can search the OS4ES Registry for DER systems that can provide the requested service and will receive a list of matching DER systems. With the help of the Planning Algorithms of his Energy Management System the aggregator can check which DER systems match best and can then reserve those DER systems for the requested time frame. This information is stored in the Contract database of the OS4ES Registry. When another aggregator searches the registry the information of this Contract database is used to evaluate if a DER system that has already been reserved for an energy service can provide another of its offered energy services without endangering the delivery of the reserved energy service. In order to achieve the above described scenario the OS4ES project will provide all necessary building blocks:

Web Based Communication Protocol: It has to be ensured that all communicating parties understand the communication protocol by means of which, e.g. DER system information is sent from the registry to an aggregator. The international standard IEC 61850 (Communication networks and systems for power utility automation) is a future-proof and the most outstanding standard for electrical networks at field and station level which also provides a standards-compliant communication framework and data model for being applied to Smart Grids. However, DER manufacturers and the industry miss a simple and low-cost web-based communication protocol in the IEC 61850 series of standards. This shortcoming has been taken up by a task force within IEC 61850 TC57 WG17 which currently evaluates web based communication protocols as alternate communication protocols to the existing IEC 61850 communication protocol MMS but only on a theoretical level. OS4ES contributes to find an apt web-based communication protocol in order to push the usage of IEC 61850 in Smart Grids. After having defined communication requirements for the OS4ES use cases it does not only evaluate web based communication protocols like OPC-UA, DPWS, XMPP, SOAP and REST on a theoretical basis - based on these requirements - at present, but will also perform practical tests based on representative network scenarios using communication network simulators. A prototype IEC 61850 web based solution will be developed for the communication protocol that turns out to be the most appropriate one among the tested ones. Results of the test simulations will be fed back to TC57 WG17. Besides, IEC 61850 protocol converters will be provided for those DER systems used in OS4ES lab and field test which only have proprietary communication protocols.

Generic Data Model for der Systems: In the same way as the communication protocol must be understood by the communicating parties it is also indispensable that all communicating parties understand the exchanged data that is transported by means of the communication protocol. Based on IEC 61850, a semantic information model for DER systems will be defined that provides the DER system data at the point of common coupling (PCC) that is relevant for the aggregator to easily and swiftly find matching DER systems for energy service requests in the registry, e.g. for setting up VPP, perform frequency control or voltage regulation. For this purpose, the information model will be so generic that it will hide the complexity of single DERs may it be generators, loads or storages - within a DER system and will provide generalized and aggregated data. Currently the OS4ES project is setting up this data model based on the use cases that have been defined in the OS4ES project. An integral part of this work is to define the Energy Services proposed in this paper. In a next step it will be analyzed in how far the part of IEC 61850 relevant for distributed energy resources, IEC 61850-7-420, needs enhancement in terms of data for the generic description of DER systems with the aim to bring in missing data in the next edition of this part of the standard. The result of this work will then be brought into the relevant standardization committees for consideration in the next edition of IEC 61850-7-420.

Registry: The registry is a core component to make a dynamic search for Energy Services and their provision possible. The distributed OS4ES registry shall bridge the gap between inventory lists for DER systems (e.g. the EEG lists in Germany) and Smart Markets for larger participants by enabling DERs to announce their availability for grid services. In addition to functional requirements such as: (a) the ability to detect matching services by defining apt filter criteria, (b) the feasibility to find single resources and (c) the availability of databases for datasheet data as well as market-related data (provided energy services, accounting-related data). The OS4ES registry consists of various core components that are designed distributed in order to meet the specified requirements:

- Index service (white pages service): a directory service of all available DER systems with nominal data and specifications as well as ICT information (e.g. communication address, voltage level).
- Yellow pages service: DER systems in the index service provide their Energy Services by the means of the Yellow pages service.

The distribution of components is based on zones that each cover parts of the power grid topology (see taxonomy in Sect. 3.2)

Middleware: A middleware is needed to allow for a seamless integration of the above mentioned components. Such a middleware will be specified and implemented in the OS4ES project. When all components (data model, communication protocol and registry) are available and are integrated in the middleware, extensive lab and field test will be conducted. Based on these results the middleware will be revised where necessary to provide a reliable and smooth central gateway for energy service provision.

5   Conclusion

In this paper, the concept of Energy Services as a generalized capability description for DERs was proposed. The concept enables the aggregator to search and find DER systems providing needed capabilities for VPP. The proposed description model handles the technical complexity of heterogeneous DER systems by providing a generic description model as Energy Services, that abstract from the technical base without leaving out necessary information for the aggregator.

The proposed Energy Service is based on the description of power in conjunction with additional information based on the DER class. One of the main research questions will be the modeling of complex DER system behavior, e.g. CHP integration with the proposed classification and description paradigm. A detailed formalization of the approach is also topic of further research efforts.

With the described Open Systems for Energy Services (OS4ES) a case study was introduced where a distributed registry system for Energy Services is currently under development. The ongoing implementation will facilitate the generalized description model of Energy Services presented in this paper to offer a registry where aggregators can search and find offered Energy Serviced based on the generalized description model. The results of the project and the field tests especially with regard to the data model implying the Energy Service definition and the registry system shall be considered in the further development of IEC 61850.

References

1. Communication networks and systems for power utility automation - Part 7–420: Basic communication structure - Distributed energy resources logical nodes. IEC Std. 61850-7-420 (2009)
2. Amin, S.M., Wollenberg, B.F.: Toward a smart grid: power delivery for the 21st century. IEEE Power Energ. Mag. **3**(5), 34–41 (2005)
3. Backers, A., Bliek, F., Broekmans, M., Groosman, C., de Heer, H., van der Laan, M., de Koning, M., Nijtmans, J., Nuygen, P., Snberg, T., Staring, B., Volkerts, M., Woittiez, E.: An introduction to the universal smart energy framework. USEF Foundation (2014)
4. Bundesnetzagentur: Smart Grid und Smart Market - Eckpunktepapier der Bundesnetzagentur zu den Aspekten des sich verändernden Energieversorgungssystems (2011)
5. CEN-CENELEC-ETSI Smart Grid Coordination Group: Sustainable processes. Brussels, Belgium, European Committee for Standardization (2012)
6. Dethlefs, T., Renz, W.: A distributed registry for service-based energy management systems. In: Industrial Electronics Society, IECON 2013, 39th Annual Conference of the IEEE, pp. 4710–4714. IEEE (2013)
7. Dielmann, K., van der Velden, A.: Virtual power plants (vpp)-a new perspective for energy generation? In: Modern Techniques and Technologies, pp. 18–20. IEEE (2003)
8. Flexible power alliance network: flexible power application infrastructure **14**, 10 (2014)
9. Kreger, H., Estefan, J.: Navigating the SOA open standards landscape around architecture. OASIS, and OMG, Joint Paper, The Open Group (2009)
10. OASIS: Universal Description, Discovery, and Integration (UDDI) 3.0.2. Organization for the Advancement of Structured Information Standards (2004)
11. Seal, B., Cleveland, F., Hefner, A.: Distributed Energy Management (DER): Advanced Power System Management Functions and Information Exchanges for Inverter-based DER Devices, Modelled in IEC 61850–90-7. Technical Report, IEC TC57 WG17 (2012)
12. Smart Grid Task Force Expert Group 3: EG3 First Year Report: Options on handling Smart Grids Data (2013)

Impact of Power-to-Gas on Cascading Failures in Interdependent Electric and Gas Networks

Andrea Antenucci, Bing Li, and Giovanni Sansavini[✉]

Reliability and Risk Engineering, Institute of Energy Technology, Department of Mechanical and Process Engineering, ETH, Zürich, Switzerland
{andrea-antenucci,libing,sansavig}@ethz.ch

Abstract. The need for an efficient and safe penetration of renewable energy resources in power systems has led to the study of unconventional storage technologies, such as Power-to-Gas (PtG). In this paper, the effects of PtG are analyzed under an operational risk perspective in the contest of interdependent gas and electrical networks. Component failures or stressful operations of one of these networks, e.g. a sudden power generation drop, can induce operational constraint violations, e.g. pressure violations, in the coupled network and lead to instability. The results show that PtG provides a beneficial impact on the pressure levels of the gas system, and entails a reduction of the consequences of a failure cascade. The beneficial effects are assessed for several configurations of PtG facilities in the gas network. Minimum safe PtG injections, which prevent pressure violations and failure cascades, are quantified.

Keywords: Power-to-Gas · Cascading failure · Critical infrastructures · Interdependent power and gas networks

1 Introduction

In the past two decades, a massive installation of renewable energy power plants has been witnessed, in particular solar panels and wind turbines, fostered by the commitment of the European Community to decarbonization [1]. However, the growing share of distributed volatile power generation rises concerns about the resilience and stability of all the related energy-carrier infrastructures. Therefore, increasing attention is devoted to the study of new and efficient ways of storing energy to cope with and fully exploit variable and volatile power generation.

Power-to-Gas (PtG) converts electrical into chemical energy under the form of hydrogen or methane, that can be conventionally stored. One of its main advantages is the possibility of using existing gas facilities or even the pipelines of the gas transmission network to store the gas produced via PtG. Several studies [2,3] address complicacies in introducing hydrogen in the natural gas network due to different physical properties, i.e. higher potential leakage rate, steel embrittlement and burners and gas turbines compatibility. Furthermore, it should be ensured that the customers are supplied within contractual obligations

© Springer International Publishing Switzerland 2015
S. Gottwalt et al. (Eds.): EI 2015, LNCS 9424, pp. 36–48, 2015.
DOI: 10.1007/978-3-319-25876-8_4

in the face of the small high heating value of hydrogen as compared to natural gas. For these reasons, the maximum achievable fraction of hydrogen that can be added to natural gas is 6 % in mass [2]. On the other hand, methane produced with PtG via the Sabatier process [4] is not affected by previous issues, although the process is less efficient. In fact, converting electricity to hydrogen entails an efficiency of 54–72 % while efficiency is 49–64 % for the conversion to methane. Globally, the efficiency of PtG is in the range 30–75 % [4].

In literature, PtG is mainly analyzed under a feasibility or economical perspective. In [4], PtG is deemed infeasible because it entails an excessive increase of the kWh price. On the other hand, [5] shows that PtG can lead to economic benefits with considerable operational costs reduction in a scenario of high share of renewable. Current research in coupled power and gas networks involves their security-constrained optimal planning [5]. The optimization can account also for random component failures [6]. Our analysis embraces a risk perspective and quantifies the consequences of several hazards which may cause operating conditions to exit the security design range. In this paper, the beneficial effects of PtG are demonstrated in the context of cascading failures between interdependent gas and electrical networks. The output of PtG facilities is directly pumped into the existing gas network; no additional reservoir is assumed in this study. Due to equipment failure or operating conditions outside the design envelope, pressure deviations in the gas network can cause the curtailment of power output in gas-fired power plants (GFPP). The resulting power redispatch may lead to the propagation of line overloads and disconnections in the electric power network, and eventually to electric instability and loss of a large share of supply. On the other hand, the gas network is also dependent on the electric power input for electricity-driven compressors, which may worsen the effects of the cascade in the electric power network. The results show that including a small share of PtG production may relieve and prevent power curtailments in the events of a localized or distributed sudden loss of power generation.

The structure of the paper is the following: in Sect. 2, a transient one-dimensional model of the gas network is presented and solved with asymmetric finite differences [7]; Sect. 3 presents the DC-power-flow-based model of the electric system which captures the propagation of line disconnections; in Sect. 4, the coupling of the two models is detailed. Section 5 presents the application to a case study consisting of interdependent electricity and gas systems, obtained combining the IEEE 39 electrical network [8] and a gas network taken from [9]. Section 6 is dedicated to the discussion of the results.

2   Gas Model

2.1   Mathematical Description of the Flow Through a Pipeline

The gas flow within a pipeline has been described via a transient one-dimensional model [7,9], which exploits the continuity equation and the law of motion:

$$\frac{\partial M}{\partial x} + S\frac{\partial \rho}{\partial t} = 0. \tag{1}$$

$$\frac{\partial P}{\partial x} + g\rho\frac{\partial h}{\partial x} + f_{\mathrm{R}} = -\rho\frac{\partial \omega}{\partial t},$$ (2)

where M = mass flow rate (kg/s); S = pipe's cross section (m^2); ρ = density (kg/m^3); P = pressure (Pa) acting on S; g = acceleration of gravity (m/s^2); h = height of the pipe element (m); ω = speed of the flow (m/s); f_R = hydraulic resistance related to unit length of the pipeline and to unit cross section.

Exploiting semi-empirical relationships for natural gas and assuming strongly sub-critical flows, (2) becomes [7]:

$$\frac{\partial \rho}{\partial x} + \frac{\lambda|\omega|}{2DS\omega_{\mathrm{sd}}^2}M + \rho\left(g\frac{\Delta h}{L\omega^2}\Big|_{\mathrm{sd}} + (1+b^*\rho)\frac{\Delta\theta}{\theta L}\right) + \frac{1}{S\omega_{\mathrm{sd}}^2}\frac{\partial M}{\partial t} = 0,$$ (3)

with λ = coefficient of hydraulic resistance; ω_{sd} = speed of sound (m/s); k = pipe's roughness (m) on the pipes diameters D (m); θ = absolute temperature (K); L = length of a pipe (m); b^* = gas constant (m^3/kg).

2.2   Integration Technique

The solution of the partial differential equations (1) and (3), has been obtained implementing the integration technique known as implicit method with intermediate step [7], which is based on asymmetrical differences and is computationally and numerically stable.

The method considers a differential equation:

$$\dot{\mathbf{y}} = f(\mathbf{y}, \mathbf{u}),$$ (4)

where $\mathbf{y}$ is the vector of the solution and $\mathbf{u}$ the vector of inputs (boundary conditions). If not linear, (4) can be linearized as:

$$\dot{\mathbf{y}} = \overline{\overline{\mathbf{A}}}\mathbf{y} + \overline{\overline{\mathbf{B}}}\mathbf{u}.$$ (5)

The sign $\overline{\overline{\cdot}}$ indicates a matrix. If Δt is the integration time step, the implicit method with intermediate step can be applied:

$$\frac{\mathbf{y} - \overline{\mathbf{y}}}{\Delta t} = \overline{\overline{\mathbf{A}}}\left((1-\theta)\mathbf{y} + \theta\overline{\mathbf{y}}\right) + \frac{1}{2}\overline{\overline{\mathbf{B}}}\Delta t(\mathbf{u} + \overline{\mathbf{u}}).$$ (6)

Dividing both sides for $1 - \theta$ ($\theta = 0.473$ as suggested by [7]), it is obtained:

$$\left(1.9\overline{\overline{\mathbf{I}}} - \overline{\overline{\mathbf{A}}}\Delta t\right)\mathbf{y} = \left(1.9\overline{\overline{\mathbf{I}}} + 0.9\overline{\overline{\mathbf{A}}}\Delta t\right)\overline{\mathbf{y}} + 1.9\overline{\overline{\mathbf{B}}}\Delta t\frac{1}{2}(\mathbf{u} + \overline{\mathbf{u}}).$$ (7)

The usage of $\theta \neq 0.5$ makes this method asymmetric and therefore different from a classical trapezoidal integration. In (6), $\overline{\mathbf{y}}$ and $\overline{\mathbf{u}}$ are computed at time t, while $\mathbf{y}$ and $\mathbf{u}$ are computed at time $t + \Delta t$. The vector $\mathbf{u}$ is sampled every time step T_{v} and, using these values, it is possible to build a function $\mathbf{u}(t)$ by linear interpolation. Defining $\Delta t = T_{\mathrm{v}}/2$, it holds:

$$\mathbf{u}(t + \Delta t) = \frac{\mathbf{u}(t + T_{\mathrm{v}}) + \overline{\mathbf{u}}(t)}{2}.$$ (8)

$$\mathbf{u}(t + \frac{T_{\mathrm{v}}}{4}) = \frac{\mathbf{u}(t + T_{\mathrm{v}}) + 3\overline{\mathbf{u}}(t)}{4}. \tag{9}$$

Therefore, the algorithm consists of the following steps:

1. Linearize the differential equation as in (5);
2. Solve the equation or the system of algebraic equations of the form:

$$\left(1.9\overline{\overline{\mathbf{I}}} - \overline{\overline{\mathbf{A}}}\Delta t\right)\mathbf{y} = \left(1.9\overline{\overline{\mathbf{I}}} + 0.9\overline{\overline{\mathbf{A}}}\Delta t\right)\overline{\mathbf{y}} + 1.9\overline{\overline{\mathbf{B}}}\Delta t\frac{1}{2}(\mathbf{u} + 3\overline{\mathbf{u}}), \tag{10}$$

   which is equal to (7) modified with $\mathbf{u}$ as defined in (8) and $\Delta t = \frac{T_{\mathrm{v}}}{2}$.
3. This is the intermediate step. Put $\mathbf{y} = \overline{\mathbf{y}}$ and solve:

$$\left(1.9\overline{\overline{\mathbf{I}}} - \overline{\overline{\mathbf{A}}}\Delta t\right)\mathbf{y} = \left(1.9\overline{\overline{\mathbf{I}}} + 0.9\overline{\overline{\mathbf{A}}}\Delta t\right)\overline{\mathbf{y}} + 1.9\overline{\overline{\mathbf{B}}}\Delta t\frac{1}{4}(3\mathbf{u} + \overline{\mathbf{u}}), \tag{11}$$

   where $\frac{1}{4}(3\mathbf{u} + \overline{\mathbf{u}})$ is found analogously to (8). $\mathbf{y}$ is the solution at time $t + \Delta t$.
4. Put $\overline{\mathbf{y}} = \mathbf{y}$ and restart from 1.

2.3   Gas Network Solution Strategy

The solutions of (1) and (3) have to be computed for all the pipes at the same time leading to complex systems of algebraic equations, which are efficiently handled as sparse matrices. Furthermore, the solution strategies have to account also for the behavior of other components, e.g. pressure regulators and compressors, which have different dynamics with respect to the pipelines. In the solution algorithm, the following terminology is used:

- Sections and nodes: a pipeline can be divided into sections, whose boundaries are called nodes. A boundary node is a node that belongs to only one section, while an internal node connects two or more sections;
- Crossing and branches: a crossing is a node common to at least 3 sections. A set of sections connecting a boundary node and a crossing, or connecting two crossings, is called a branch;
- Non-pipe element: is a fictitious branch that represents a particular element, such as a compressor or a pressure governor.

Solving the gas network means to find densities ρ and mass flows M at each section of each pipe. However, since ρ and M are not independent, from (1) and (3) it is possible to express the latter, e.g. for an internal branch, as:

$$M_i = A + B\rho_1 - C\rho_{n+1}, \tag{12}$$

where M_i is mass flow in branch i, A, B and C are coefficients obtained by the rearrangement of (1) and (3), while ρ_1 and ρ_{n+1} are respectively the densities in the first and last section of branch i. For each crossing, the continuity equation is written with ρ as the only unknown and solved. The matrix obtained by this set of equations is found to be sparse and therefore its computational cost is low. Setting the computed density values as boundary conditions for the branches, the pair (ρ, M) can be computed in all the sections, exploiting the continuity and motion equations. Again, the system of algebraic equations for each branch leads to sparse matrices. This system is different for internal and external branches, because they require different boundary conditions.

The introduction of non-pipe elements implies a change in the algorithm when computing the densities and mass flows for these elements. Their characteristic equations and the state of the system determine the pair (ρ, M).

The modelling of the gas network required some simplifications. λ is considered time-independent and constant throughout the network, as commonly exploited in literature [7,9]. Moreover, the inlet of the network is modelled via a pressure governor. PtG production is modelled as a constant injection of gas through one or more nodes of the gas network. The gas network model has been validated through comparison with a test system in [9].

2.4   Operational Constraints in the Gas Transmission Network

The gas network is subjected to several constraints, defined in order to represent physical, operational and contractual limitations. Pressure fluctuations are allowed in the network within an operative range. Maximum pressure limits are established according to the material resistance of the pipelines. Minimum pressure values are chosen independently for each node, and represent the minimum contractual pressure to be provided at each delivery point, in order to guarantee the normal operations of power plants and compressors. Compressor stations are bound to remain within an operative mass flow and pressure ratio envelope. The compressor operative point defines the amount of electrical power required from the electrical network:

$$P_{\text{compressor}} = \frac{P_{\text{suction}} * Q}{\eta * m} * [\beta^m - 1],\qquad(13)$$

where $P_{\text{suction}} =$ pressure at the compressor inlet (Pa), $Q =$ volumic mass flow through the compressor (m^3/s), $\eta =$ compressor efficiency, $m =$ gas constant and $\beta =$ pressure ratio. The out-take rate is limited by the ramp rates of the power plant generators [10].

The violation of one or more constraints entails a counter-action from the system, in order to remain within safety thresholds. The corrective action is specific for each constraint typology. Minimum pressure violation is relieved by gas curtailment at the location of the violation. In absence of gas off-take, the curtailment action takes place at a proximate node. In this model, the curtailment strategy is based on five pressure limits; every time the pressure goes below the threshold values, further curtailment occurs at the power plant, until the pressure starts raising again. Ultimately, the entire power plant is shut down. The amount of power reduction and the pressure limits depend on the production strategy and the technical characteristics of each power plant.

Compressor violations can be caused by the lack of electrical power provided to the machines or by surge. Upon violations, compressors are shut down. Mass flow passing through the failed element decreases following a transient that lasts 3 min. In order to avoid reverse mass flow, gas flow is prevented until the pressures at the inlet and outlet of the compressor are equal. Table 1 summarizes the effects and the corrective actions following each constraint violation.

Table 1. Constraint limitations and corrective actions in the gas network.

Constraint	Value	Effect/Correction
Maximum pressure	100 bar	Gas curtailment
Minimum pressure	10–42 bar	Gas curtailment
Compressor envelope	4000–9500 rpm	Compressor shutdown
Compressor envelope	3500–20000 m^3/h	Compressor shutdown
Power required by the compressor	Depending on working set point	Compressor shutdown
Ramp rate of PP	0.5 p.u./h	Ramp up/down limits

3   Electrical Model

The electric model employs a DC power flow in order to study the system behavior under failure conditions [11]. Disruptive contingencies, such as line disconnections, power plant (PP) failures or compressors shut down can be introduced. The model evaluates power flow changes and simulates redispatch and load shedding. Furthermore, line overheating is computed exploiting a transient temperature model, that accounts for the actual power on the line and the line typology and characteristics [8]. This model assumes that additional random failures cannot occur during the cascade event, given their low probability.

The electrical model is characterized by an event-base structure, where 4 types of events are defined. Therefore, the simulation does not evolve through a fix time-step, but moves forward in time from one event to the following. The events are:

1. Time to next hour: the model evaluates the state of the system at each hour, sampling the external electricity demand;
2. Line disconnection: based on the power flowing within the lines, the model predicts the evolution of the line temperatures and detects when to disconnect a line, due to overheating protection;
3. Constraint violation in the gas network: when there is a violation in the gas network, a new event is created in the schedule of the electrical network. This allows the two models to communicate with each other. The dynamic of this process will be further explained in Sect. 4;
4. Time to power balance restoration: given an imbalance between load and generation and the ramp rates of generators, the model computes the time when power balance is restored. Power redispatch is performed through primary frequency control.

3.1 Operational Constraints in the Electric Network

Operational constraints in the electrical network concern maximum power imbalance, lines and generation characteristics. Maximum frequency deviation is set at 5 % of the base frequency of the system, which is 50 Hz. Frequency deviation is computed as:

$$f.d. = \frac{\Delta P}{\sum_{\Omega_D} D_\mathrm{d} + \sum_{\Omega_G} \frac{1}{R_\mathrm{g}}}, \tag{14}$$

with ΔP = power imbalance, D_d = frequency characteristics of the $\mathrm{d^{th}}$ load (MW/Hz), R_g = frequency characteristics of the $\mathrm{g^{th}}$ generator (Hz/MW), $\sum_{\Omega_D}$ = set of demands and $\sum_{\Omega_G}$ = set of generators [12]. Power plants are characterized by maximum and minimum power output, and max ramp up/down limitations. Finally, power lines are constrained by the amount of power that can pass through them, and by the maximum temperature during normal operations, which is set to 100 °C. Table 2 summarizes the constraints and the resulting effects or corrective actions.

Table 2. Constraint limitations and corrective actions in the electric network.

Constraint	Value	Effect/Correction
Frequency deviation	5 %	Island blackout
Maximum generation	680–1100 MW	Generation limitation
Minimum generation	340–550 MW	Power plant shutdown
Line flow limit	707 MW	Line disconnection
Ramp rate of PP	0.5 p.u./h	Ramp up/down limits

4 Interdependent Electric/Gas Model Description

The coupling between the gas and the electric systems takes place via compressors and power plants. The relation that links gas out-takes to power generation is:

$$\text{Electric power supplied by a GFPP} = M * HHV * \eta, \tag{15}$$

where M = off-take mass flow (kg/s), HHV = higher heating value of natural gas (J/kg) and η = overall GFPP efficiency.

The simulation is initialized with gas and electrical networks at steady state. The event which could possibly generate a failure occurs at time t* (e.g. line trip(s) in the electrical network, power plant outage(s), change in power generation profile).

The following steps are executed:

1. The next event is selected, topology modifications are applied to the networks and, if electric power generation and consumption are balanced, network flows are updated;

2. The time intervals Δt_i to the potential occurrence of events 1–3 (Sect. 3) are updated;
3. Island identification in the electrical system and power redispatch in each island are performed. Load shedding occurs if frequency deviation is larger than 2.5 Hz (5 % of 50 Hz). The time interval Δt to event 4 (Sect. 3) is updated;
4. The gas model is executed using the boundary conditions given by the electrical network till the minimum Δt_i, namely $\Delta t_{\text{next event}}$, is reached. However, if a gas constraint violation occurs at $\Delta t_{\text{violation}} < \Delta t_{\text{next event}}$ the gas simulation stops at $\Delta t_{\text{violation}}$. Gas off-take curtailment is applied and the electric power output from GFPPs is updated. An event 3 (constraint violation in the gas network) is created in the electrical network schedule at $\Delta t_{\text{violation}}$. If no violation occurs before Δt_i, the output of the gas nodes is able to supply the request of the electrical nodes and $\Delta t_{\text{violation}} = \infty$;
5. The minimum time interval Δt_i identifies the next event to occur.

Steps 1–5 are repeated until the simulation time is reached.

Gas and electrical demands are boundary conditions of the coupled networks. PtG is also a boundary condition, and the amount of gas produced via PtG is quantified as a percentage of the total gas demand. This is equivalent to assuming that the gas is produced during high peaks renewable power production and stored in the proximity of PtG facilities.

5   Case Study

This paper analyzes the behavior of an interdependent gas and electrical system obtained combining the IEEE 39 power network [8], and a gas network consisting of 40 branches and 27 nodes [9]. A schematic representation of the two systems is given in Fig. 1.

Interdependencies are identified by common elements, i.e. 9 GFPPs and 3 electricity-driven compressors. Additional electrical generation includes one hydro power plant. The main gas injection point is indicated by the pressure governor, marked as 1. PtG facilities are not explicitly illustrated because their positions vary in different scenarios as illustrated in Sect. 6.

In Sect. 6, the effects of PtG on the system safety are shown through two exemplary cases. PtG mitigating effects are evaluated analyzing the system response to random failures, i.e. power plant shut down, in terms of demand not served (DNS). DNS is computed as the difference between the power demand and the power supplied to the electric loads. Furthermore, PtG effects are analyzed when GFPPs are used to compensate for a sudden loss of generation stemming from a negative ramp in renewable energy conversion plants. A sudden interruption of renewable power generation might stress the gas network towards critical working conditions, i.e. increasing off-takes may lead to minimum pressure violations.

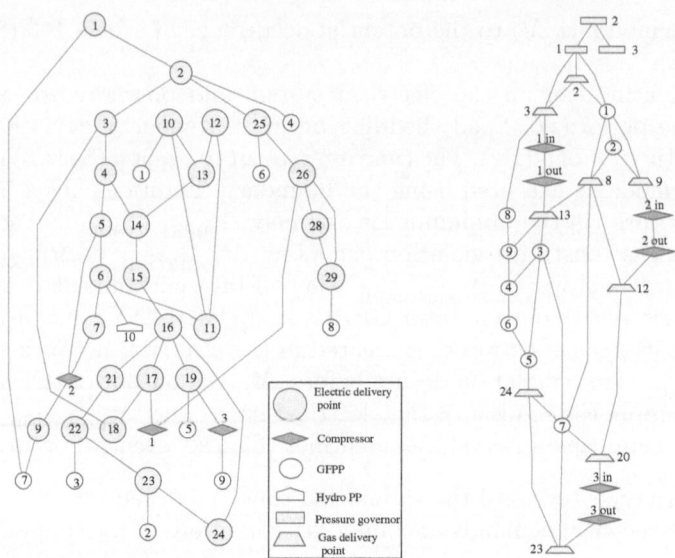

Fig. 1. Electrical (left) and gas network (right). The legend shows which symbol is used for each element type. The elements are sequentially numbered for each typology.

6 Results and Discussion

6.1 Effect of PtG Following a Random Component Failure

The initiating events concerning the shutdown of one gas-fired power plant are simulated. The simulation time covers the 25 h following the initial event. By this time, the two networks have reached a new stable state and the consequences of the various scenarios can be compared. Increasing amounts of gas injected via PtG facilities are analyzed following the initial event. The total amount of renewable gas injected is quantified as a share of the total gas demand (440 kg/s) and it ranges from 0 to 5 %. The effects of increasing the share of PtG are evaluated in terms of DNS computed at the end of the simulation.

In order to uncover possible topological effects stemming from the different siting of PtG injection points, the position of the PtG facilities can vary. To this aim, for each initial event and PtG share, three PtG configurations are assessed representing distributed and localized siting, namely, (1) distributed PtG production facilities uniformly located at 25 gas nodes; (2) PtG localized at the node with maximum gas demand, i.e. gas node 23; (3) PtG localized at a peripheral node with small gas demand, i.e. gas node 19 (marked as PP5 in Fig. 1, right).

The networks are initialized at steady-state conditions, computed with the following levels of external demands, namely, 6254 MW of electrical power demand and 257 kg/s of non-electrical gas demand (58 % of the gas in the system). This steady state represents stress operations for the coupled networks,

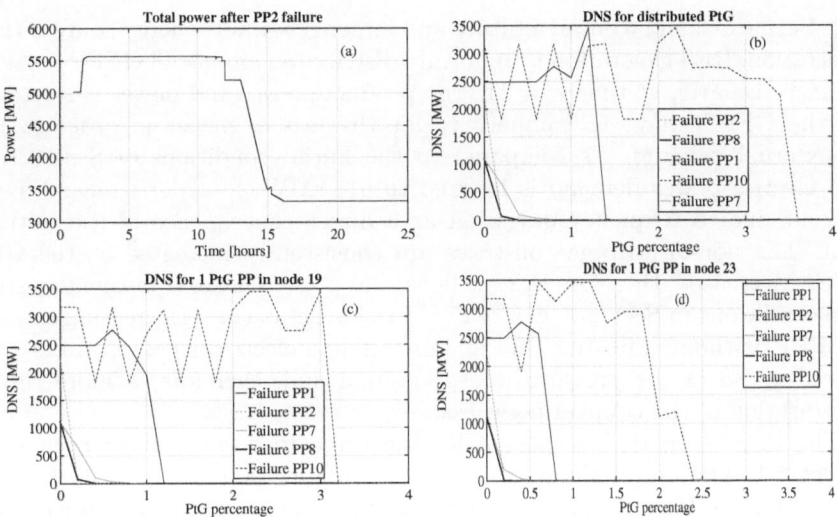

Fig. 2. Example of cascading failures on the *total power generation (a)*. Values of *DNS* for increasing *PtG percentages*, for configurations: distributed *(b)*, in node 19 *(c)*, in node 23 *(d)*.

and it is representative of a very cold winter day, when high electrical consumptions are coupled with a large gas demand for heating. At steady state, the initial pressure at each node is assumed to be 5 % larger than the minimum contractual pressure. This value is selected by network operators as a tradeoff between minimum pressure safety margin and operational costs for compressor stations.

In Fig. 2a, an example of cascading failure on the total power generation during the simulation time in shown. In Fig. 2b–d, the DNS registered at the end of the simulation time is illustrated for an increasing PtG share in scenarios 1–3. For all the configurations, a PtG share around 4 % is sufficient to inhibit all the effects of the cascade. Localized distributions (Fig. 2c and d) induce larger improvements to the resilience of the system, given that less PtG injection is needed for matching the same results of the distributed case. The DNS oscillations for increasing PtG injections, e.g. for the initial failure of PP10 in Fig. 2b–d, result from the formation of different islands at the end of the cascade. In this case, the propagation of the cascading failure shows that PtG injections increase pressures at some nodes, hindering the cascading process. This changes the chain of events and the following island formations. These oscillations demonstrate a strong non-linear behavior of the coupled networks.

6.2   Effect of PtG Following a Sudden Lack of Renewable Power

In order to evaluate the effect of a sudden drop in renewable power production, the generation mix has been rearranged in three different scenarios, summarized in Table 3. Scenario 1 is coherent with the wind power fluctuations analyzed in [13], while Scenario 2 and 3 entail a future and larger renewable energy sources (RES) penetration. RES generation is uniformly distributed among all GFPP locations. For each scenario, at time $t^* = 2\,h$, the production of wind power is set to zero and the GFPPs have to compensate for the loss of power in order to meet the external demand. As compared to the initial conditions of Sect. 6.1, the total electric power demand is lowered to $5164\,MW$, in order to avoid system blackout due to frequency deviation as a direct consequence of the initiating event. The non-electric gas off-takes are consistently increased as the GFPP share is decreased, therefore the gas flows and pressures are equivalent to steady state conditions in Sect. 6.1, before the initial event occurs. Decreasing the share of GFPP without adjusting the amount of non-electrical gas off-takes would lead to a rise of the pressures of the system and, therefore, a more resilient configuration of the coupled networks.

The three scenarios are tested for the three PtG configurations described in Sect. 6.1. The only difference is that in configuration (3) the PtG facility is placed at node 22, i.e. the inlet of the compressor 3, which has proven to be susceptible to minimum pressure violations.

Figure 3 shows the minimum share of PtG (relative to a consumption of $440\,kg/s$) that prevents the outbreak of a cascading failure, i.e. DNS = 0 during the entire simulation time, for the three PtG configurations and for the' three power generation portfolios in Table 3. This PtG share is different from what is presented in Fig. 2b–d, which entails a zero DNS at the end of the cascading failure. Therefore, the PtG share shown in Fig. 3 are larger than the PtG share which entails DNS = 0 in Fig. 2b–d. For all the three PtG configurations, a larger drop in renewable power generation requires increasing amounts of PtG injections in order to avoid pressure violations in the gas network. Furthermore, localized gas injections in strategic points of the gas system, e.g. in node 22, enhance the PtG beneficial effects, even though the improvements amount at a

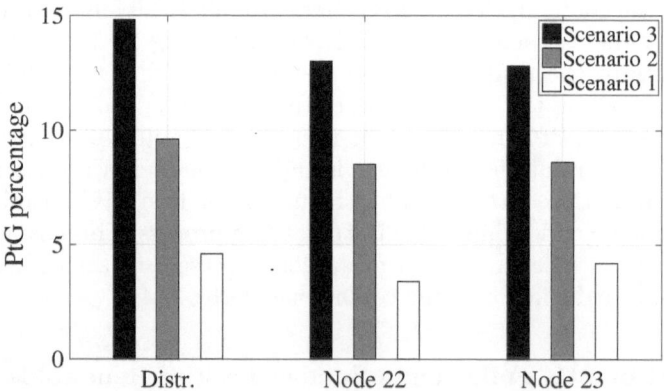

Fig. 3. *PtG percentage* for 3 PtG configurations and three generations portfolios in Table 3.

Table 3. Simulation scenarios involving three generation portfolios.

Scenario	Wind power share	GFPP share	Hydro PP share
1	10 %	85 %	5 %
2	20 %	75 %	5 %
3	30 %	65 %	5 %

few percentage points as compared to the distributed configuration. For Scenario 1, which is representative of a realistic penetration of RES, including roughly 4 % of distributed PtG production prevents pressure violations and cascading failures. This amount of PtG injection would require 25 PtG plants of around 80 MW each, considering a global facility efficiency of 52 %.

7   Conclusions

In this paper, the effects of PtG on cascading failures in interdependent electric and gas networks have been assessed. The results show that PtG can positively affect the security of the interdependent critical infrastructures analyzed, via the enhancement of the pressure levels of the gas infrastructure. Depending on the extent of the initiating failure and for a network of the considered size, a relatively small share of gas injected via PtG can prevent pressure violations and cascading failures triggered by the random failure of a power plant. Moreover, a small share of PtG injection can compensate for large power generation drops in scenarios with 10 % RES penetration and guarantee reliable operations of the coupled networks. The beneficial effects are larger for a localized PtG configuration than for a distributed configuration, even though the total amount of gas required to prevent cascading failure in both PtG configurations is similar.

As a possible improvement, an AC electrical model could be implemented. This may lead to different values of DNS due to considering the voltage profile and the reactive power demand in the cascading failure process.

References

1. European Council Conclusions (EUCO 169/14) (2014)
2. Tabkhi, F., Azzaro-Pantel, C., Pibouleau, L., Domenech, S.: A mathematical framework for modelling and evaluating natural gas pipeline networks under hydrogen injection. Int. J. Hydrogen Energy **33**(21), 6222–6231 (2008)
3. Gahleitner, G.: Hydrogen from renewable electricity: an international review of power-to-gas pilot plants for stationary applications. Int. J. Hydrogen Energy **38**(5), 2039–2061 (2013)
4. Schiebahn, S., Grube, T., Robinius, M., Zhao, L, Otto, A., Kumar, B., Weber, M., Stolten, D.: Power to gas. In: Transition to Renewable Energy Systems, 1st edn., pp. 813–848 (2013)

5. Qadrdan, M., Abeysekera, M., Chaudry, M., Wu, J., Jenkins, N.: Role of power-to-gas in an integrated gas and electricity system in Great Britain. Int. J. Hydrogen Energy **40**(17), 5763–5775 (2015)
6. Chaudry, M., Wu, J., Jenkins, N.: A sequential Monte Carlo model of the combined GB gas and electricity network. Energy Policy **62**, 473–483 (2013)
7. Krlik, J., Stiegler, P., Vostry, Z., Zavorka, J.: Dynamic Modeling of Large-Scale Networks With Application to Gas Distribution. Elsevier, Amsterdam (1988)
8. IEEE 39-Bus System - Illinois Center for a Smarter Electric Grid (ICSEG). http://publish.illinois.edu/smartergrid/ieee-39-bus-system/
9. Osiadacz, A.J.: Simulation and Analysis of Gas Networks. Gulf Publishing Company, Houston (1986)
10. European Commission: Study on Synergies between Electricity and Gas Balancing Markets (EGEBS) (2012)
11. Sansavini, G., Piccinelli, R., Golea, L.R., Zio, E.: A stochastic framework for uncertainty analysis in electric power transmission systems with wind generation. Renew. Energy **64**, 71–81 (2014)
12. Mousavi, O.A., Bozorg, M., Cherkaoui, R., Paolone, M.: Interarea frequency control reserve assessment regarding dynamics of cascading outages and blackouts. Electr. Power Syst. Res. **107**, 144–152 (2014)
13. Kamath, C.: Understanding wind ramp events through analysis of historical data. In: 2010 IEEE PES Transmission and Distribution Conference and Exposition: Smart Solutions for a Changing World (2010)

Smart Meters and Monitoring

The RASSA Initiative – Defining a Reference Architecture for Secure Smart Grids in Austria

Marcus Meisel[1](✉), Angela Berger[2], Lucie Langer[3], Markus Litzlbauer[1],
and Georg Kienesberger[1]

[1] Technische Universität Wien, Gußhausstaße 27-29, 1040 Vienna, Austria
{meisel,kienesberger}@ict.tuwien.ac.at, litzlbauer@ea.tuwien.ac.at
[2] Technologieplattform Smart Grids Austria,
Mariahilferstr. 37-39, 1060 Vienna, Austria
angela.berger@smartgrids.at
[3] Digital Safety and Security Department, AIT Austrian Institute of Technology
GmbH, Donau-City-Straße 1, 1220 Vienna, Austria
lucie.langer@ait.ac.at

Abstract. The goal of the *Reference Architecture for Secure Smart Grids in Austria (RASSA)* initiative is to design and establish a technical reference architecture specification in coordination with all relevant stakeholders. This goal is realized across multiple projects. This paper first motivates the need for developing a coordinated smart grids reference architecture for Austria involving all relevant actors, such as infrastructure operators, manufacturers, and public agencies. After a description of most prominent international reference architecture efforts, first results on how to develop a reference architecture serving as a blueprint for further smart grids solutions is described. Necessary coordination and communication efforts to achieve a nationally accepted and internationally aligned process are described. The paper closes with an outlook on a practical application of the principles defined in order to meet stakeholder requirements through target-group-specific involvement.

1 Motivation

Global electricity systems are undergoing a radical change. In the course of intensive efforts to raise the share of renewable energy sources, new innovative smart grids solutions have been developed in the past years in order to integrate decentralized volatile generation. With the introduction of smart grid technologies, an interconnection with communication technologies has taken place, changing the accessibility of previously isolated assets especially in the distribution grid. This has lead to challenges in system design in terms of cybersecurity, interoperability, and security of supply.

Over the past years, the necessity of a holistic approach to achieve a national secure smart grid reference architecture, dealing with critical requirements not addressed by European standardization organizations or the Smart Grids Reference Group, was frequently discussed within the Technology Platform Smart

© Springer International Publishing Switzerland 2015
S. Gottwalt et al. (Eds.): EI 2015, LNCS 9424, pp. 51–58, 2015.
DOI: 10.1007/978-3-319-25876-8_5

Grids Austria. As a consequence, the Platform launched the RASSA initiative as its core undertaking. The objective of the RASSA initiative is to develop a national reference architecture for smart grids, building on European and international activities and considering stakeholder needs. Aspects like operational safety, cybersecurity, and privacy are considered throughout the design and evaluation of the architecture ("by design"). A successful development of a reference architecture is only possible if all relevant stakeholders, like network operators, energy suppliers, regulators, and public agencies, are involved from the very beginning. This is challenging as the number of actors is high due to the significant economic relevance and criticality of the energy supply system.

The comprehensive stakeholder process is therefore a core part of the RASSA initiative. Experts and decision makers from the energy sector, manufacturing industries and public bodies are closely involved. This ensures that a solution for the specific Austrian needs is found, while taking into account Austria's situation within the European network as well as maintaining full compatibility with international standards.

While the *RASSA Architecture* project focuses on the development of the Austrian reference architecture, the *RASSA Process* project aims at providing methodological support and establishing a sustainable stakeholder involvement. The stakeholder process builds upon existing good contacts of the coordinating Technology Platform Smart Grids Austria, as its members are part of the relevant stakeholder groups. Furthermore, the Platform has developed a solid basis for discussion with actors outside the energy sector, such as ministries, public agencies, and international players. The designed stakeholder concept has to accompany the whole development process of smart grids in Austria and must therefore be sustainable. As relevant preparatory work the Technology Platform Smart Grids Austria developed a Technology Roadmap for Smart Grids in the last year [1], in which a broad agreement of the energy sector regarding the next necessary steps for the transformation of the energy system towards smart grids was expressed. One of the key steps identified was the development of an overall ICT architecture for smart grids. Based on these results, the development of the Austrian reference architecture has a great chance to become widely accepted from a national point of view.

2   State of the Art

This section summarizes existing international work which will be drawn upon for the design of the Austrian reference architecture. Although plenty of material exists in this regard, the migration path from existing to future grid implementations is rather unclear, and has to be defined on a national basis. The collection of national efforts to create a smart grid landscape for Austria as a basis for the migration path is still ongoing; early results are presented in Sect. 4 of this paper.

2.1   The Smart Grid Architecture Model (SGAM)

The widely-accepted Smart Grid Architecture Model (SGAM) has been defined as part of CEN-CENENELEC-ETSI's response [2] to the EU Smart Grid Mandate M/490. SGAM is defined by *zones, domains*, and *interoperability layers* (see Fig. 1). While the zones are derived from the hierarchical levels of information management in power systems (from field via process, station towards operation, and enterprise level), the domains reflect the different stages of power generation, transmission, distribution, and consumption within the electrical energy conversion chain. Electrical domains and information management zones span the *smart grid plane*. In the third dimension, SGAM features five interoperability layers which are an abstracted and condensed version of the originally eight GridWise interoperability layers [3], and represent different stakeholders' views. The base layer is the component layer, which represents physical devices and software components. On top of that, communication protocols and mechanisms for the exchange of information between different components are represented in the communication layer. The information layer represents information objects or data models required to fulfill functions and to be exchanged by communication. The function layer represents logical functions or applications independent from physical implementations, and the uppermost business layer describes business models and regulatory requirements.

SGAM has proven useful for describing use cases within a given European grid, establishing a common view between different stakeholders. The Austrian reference architecture under development will also be mapped to SGAM, which requires specifying the national particularities on each interoperability level.

2.2   BSI Protection Profiles

Germany's Federal Office for Information Security (BSI) has developed a Common Criteria Protection Profile for the Gateway of a Smart Metering System and its Security Module [4,5]. Based on a threat analysis, both profiles define a set of minimum security requirements. While these documents are important input to the reference architecture definition in terms of security and privacy aspects for smart meter devices, a different legal standing applies within Austria. Furthermore, additional parts of a smart grid have to be considered.

2.3   NIST Reports and Frameworks

The U.S. National Institute of Standards and Technology (NIST) developed the Roadmap for Smart Grid Interoperability Standard (NIST-SR 1108R3) [6] and a multitude of reports on smart grids, such as the Guidelines for Smart Grid Cyber Security (NIST-IR 7628) [7]. This report identifies seven smart grid domains and a logical interface architecture used to define categories of interfaces within and across those seven domains. The security requirements for these interface categories are identified through a risk assessment process, which relies on a top-down and a bottom-up approach. Whilst the top-down approach defines

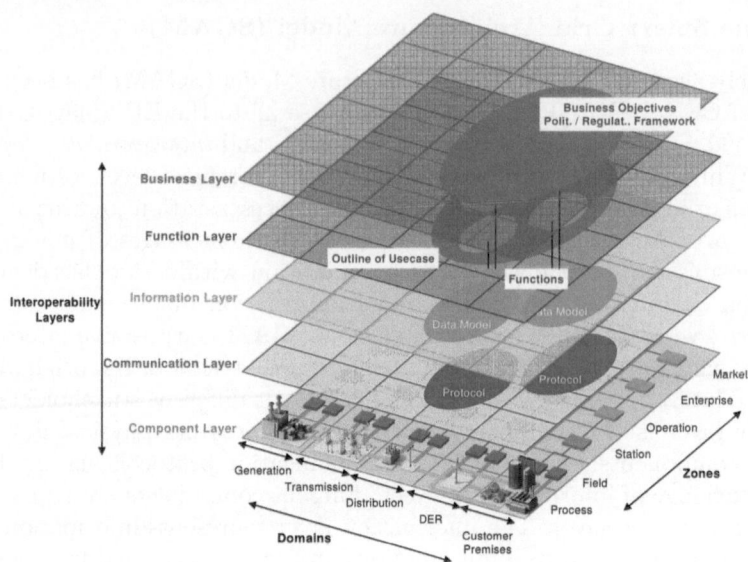

Fig. 1. Smart grid architecture model by CEN-CENELEC-ETSI [2]

smart grid components and interfaces, the bottom-up approach focuses on cyber-security issues in power grids, such as user authentication, key management for meters, and intrusion detection for power equipment. NIST-IR 7628 can be used to design and integrate smart grid technologies and is therefore a useful tool in the reference architecture development process.

3 RASSA Initiative Overview

The related work shows that international concepts are vital for highly inter-connected smart grids. However, national peculiarities cannot be sustainably handled at that level and need to be broken down for a more detailed analysis, still without dictating a single specific solution on component or communica-tion level. Instead, a holistic approach is required, which takes into account requirements from the institutional framework, which maps societal goals to leg-islative or regulatory framework conditions. The goals of the RASSA initiative can therefore be summarized as follows:

1. Define a consistent reference architecture for the Austrian smart grid, thor-oughly considering aspects like security, safety, resilience, privacy,
2. Reach a consensus among all relevant stakeholders on contents and utilization of the reference architecture,
3. Foresee sufficient degrees of freedom to allow a competitive realization of the suggested reference architecture, and
4. Provide concrete guidance on the migration path from today's grid to a future smart grid infrastructure as expressed by the reference architecture.

The institutional framework has implications for business models and enterprise processes regarding how the various players interact and organize their business. The processes defined eventually affect the necessary information exchange and data models. In addition to this, the architecture solution of course has to consider available technology. The reference architecture development cannot simply build on a one-time acquisition of requirements, but rather is an iterative process which is continuously adapted to the ever-changing challenges of technology development.

The first project within this initiative, *RASSA Process*, is mainly concerned with establishing concepts for continuous stakeholder involvement. Details are described in Sect. 4. One of the result of the stakeholder process is a prioritization of topics in terms of high-level use cases. In addition to concept development, the project deals with the technical and scientific fundamentals of reference architecture development.

While the *RASSA Process* project aims at providing methodological support and establishing a sustainable stakeholder involvement, the *RASSA Architecture* project focuses on developing the core reference architecture (see Sect. 5). Figure 2 shows the relationship of the two projects and their thematic priorities. Follow-up projects within the RASSA initiative will focus on different aspects such as risk management for smart grids.

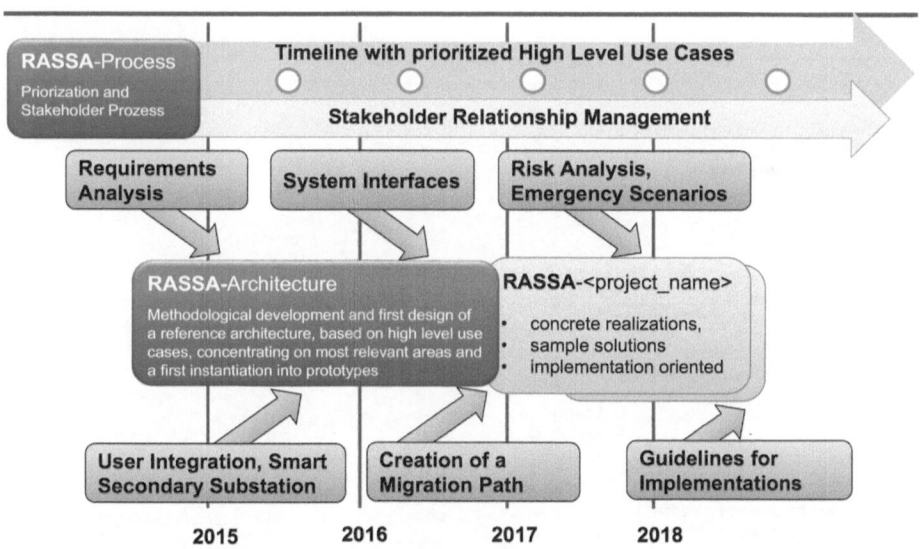

Fig. 2. Timeline of the projects part of the RASSA initiative

4   Preliminary Results of RASSA Process

A special feature of the RASSA initiative is its support by a broad institutional base, namely the partners of Technology Platform Smart Grids Austria from energy industry, regulation, technology, and research. However, the challenge of RASSA cannot be mastered by these actors alone, but requires coordination between all relevant stakeholders on different technical and organizational levels. Due to the large number of stakeholders only a structured approach can achieve wide acceptance of the solution. A key objective of the project is to clarify the working structures and interfaces of the different stakeholders. The work will build on previous activities in Austria and existing international models like SGAM and the NIST proposals, but will also extend beyond the boundaries of energy, communication and information systems.

4.1   Analyze Stakeholders

Beginning with *RASSA Process*, relevant stakeholders and their demands in relation to the energy infrastructure were identified. In Austria, responsibilities in the smart grids area such as energy, research, safety, and economy are divided among several ministries and agencies. Therefore, an integrated strategy development is difficult, but a coordinated approach is essential – especially in areas of critical infrastructure and security. It is now time to carry out this coordination within affected areas, in order to create a common strategy framework for industry and energy. Figure 3 illustrates different stakeholders and their diverse requirements regarding the energy infrastructure.

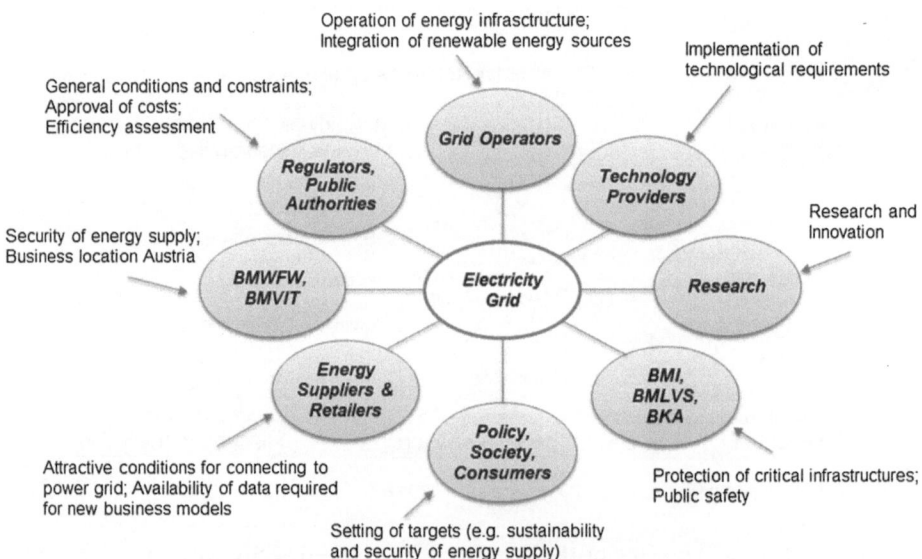

Fig. 3. Stakeholders and their requirements

4.2   Organize, Prioritize, and Contact Stakeholders

As a next step, the stakeholders were prioritized according to their impact on the development of the reference architecture. Currently, they are being contacted following target-group-specific approaches. The focus is on creating a common vision and to foster the willingness to work together. A stakeholder workshop has been held at Smart Grids Week in May 2015 to identify the existing knowledge gap and suitable scientific methods for the development of a reference architecture. Aspects considered were, for example:

- Where the Austrian reference architecture will differ from other countries,
- International regulatory issues and other constraints,
- Whether an Austrian reference architecture can serve as template for other countries, and
- Relevant use cases and methods for developing a reference architecture.

4.3   Define Concept for Stakeholder Involvement

As a final step, an information exchange concept will be defined to guarantee active involvement of stakeholders throughout the architecture development process. A coordination of stakeholder requirements and their consideration in architecture development is currently planned for *RASSA Architecture* and following projects (see Sect. 5). In addition to national activities, an international coordination is intended. Experience gained from European development processes, for example, in M/490 is provided through a subcontract with OFFIS. Workshops with international participation were already held, and especially the know-how exchange within the D-A-CH region is actively being pursued.

5   RASSA Architecture Outlook

The *RASSA Architecture* project started in June 2015 and runs till November 2017. Its aim is to develop a consistent and unifying reference architecture for smart grids in Austria, and establish its acceptance among all relevant stakeholders such as grid operators, energy suppliers, regulators, and public authorities. Based on existing standards and concepts including SGAM (see Sect. 2), the project will specify a reference architecture for Austria that can be used as a blueprint for smart grid implementations. By instantiating parts of the reference architecture within the boundaries given by its parameters, secure and interoperable smart grid solutions can be developed in a straightforward and consistent way. The reference architecture will not be unnecessarily prescriptive; sufficient degrees of freedom will allow grid operators and manufacturers to make their own decisions how to instantiate different components. This will allow innovation to continue in this important emerging market. The practical applicability of the reference architecture will be demonstrated in the areas of an innovative smart secondary substation and in the customer domain. This will be achieved by applying the conditions defined by the architecture to existing components that are provided by grid operators and manufacturers within the consortium. Special consideration will be given to customers by designing and integrating privacy-enhancing technologies (PETs) into the reference architecture.

A harmonized reference architecture for Austria, which is synchronized with Europe and further afield, will strengthen the competitiveness of Austrian companies and research organizations, by enhancing the Austrian market to become a pilot market with clear technical requirements.

Acknowledgment. This paper is based on findings of the project *Initiative Referenzarchitektur Sichere Smart Grids Austria – Projekt RASSA-Prozess*, which was commissioned by the Austrian Climate and Energy Fund as part of the *1st Call Stadt der Zukunft* in the topic *Smart Grids Reference Architecture*.

References

1. Technologieplattform Smart Grids Austria (TPSGA): Technologieroadmap Smart Grids Austria - Die Umsetzungsschritte zum Wandel des Stromsystems bis 2020. Technical report, Technologieplattform Smart Grids Austria, April 2015. http://www.smartgrids.at/index.php?download=372.pdf
2. Smart Grid Coordination Group: Smart grid reference architecture. Technical report, CEN-CENELEC-ETSI, November 2012. http://www.cencenelec.eu/standards/Sectors/SustainableEnergy/SmartGrids/Pages/default.aspx
3. GridWise Architecture Council Interoperability Framework Team: Interoperability Context-Setting Framework. Technical report, GridWise Architecture Council, July 2007. http://www.caba.org/resources/Documents/IS-2008-30.pdf
4. Kreutzmann, H., Vollmer, S.: Protection profile for the gateway of a smart metering system (smart meter gateway pp). Technical report BSI-CC-PP-0073, Bundesamt für Sicherheit in der Informationstechnik (BSI) Federal Office for Information Security, Germany, March 2014. https://www.bsi.bund.de/DE/Themen/SmartMeter/Schutzprofil_Gateway/schutzprofil_smart_meter_gateway_node.html
5. Federal Office for Information Security Germany: Protection profile for the security module of a smart meter gateway (security module pp). Technical report BSI-CC-PP-0077-V2, Bundesamt für Sicherheit in der Informationstechnik (BSI), December 2014. https://www.bsi.bund.de/DE/Themen/SmartMeter/Schutzprofil_Security/security_module_node.html
6. Smart Grid and Cyber-Physical Systems Program Office and Energy and Environment Division, Engineering Laboratory, Physical Measurement Laboratory, Information Technology Laboratory: NIST Framework and Roadmap for Smart Grid Interoperability Standards Release 3.0. Technical report SP 1108R3, NIST, February 2014. http://www.nist.gov/smartgrid/upload/NISTDraftFrameworkOct_2013.pdf
7. Smart Grid Interoperability Panel: Guidelines for smart grid cyber security. Technical report 7628, Cyber Security Working Group, (NIST), September 2010. http://www.nist.gov/smartgrid/upload/nistir-7628_total.pdf

NoFaRe: A Non-Intrusive Facility Resource Monitoring System

Matthias Kahl[✉], Christoph Goebel, Anwar Ul Haq, Thomas Kriechbaumer, and Hans-Arno Jacobsen

Technische Universität München (TUM), Munich, Germany
matthias.kahl@in.tum.de

Abstract. The aim of this paper is to present the idea and starting points for an innovative facility resource monitoring system, which will be realized in a recently started research project: NoFaRe. NoFaRe's goal is to enable low cost monitoring of electrical devices in buildings using advanced Non-Intrusive Load Monitoring (NILM) techniques and evaluate its value in facility management based on Building Management System (BMS) prototypes. Low-level device monitoring in buildings is a necessary first step to realize a new generation of BMS that will allow for higher service and efficiency levels in various dimensions of facility management. The general goal of NILM algorithms is to obtain information on the behavior of single appliances based on aggregate measurements, such as smart metering data, which allows for reducing the required amount of sensors and communication infrastructure. The NoFaRe project will on the one hand explore innovative NILM concepts to fulfill BMS application requirements while minimizing hardware cost. On the other hand, it will contribute innovative BMS applications based on device-level monitoring and contemporary communication infrastructure.

Keywords: Electricity metering · Building Management Systems · Non-Intrusive Load Monitoring

1 Introduction

Modern information technology is about to fundamentally change the way facilities are managed. Many buildings already incorporate increasingly sophisticated Building Management Systems (BMS) that integrate building control with improved sensors and better data collection and presentation capabilities. However, these systems currently only allow for simple, decoupled control of building services, such as lighting, ventilation, heating, and cooling. Their architecture and Application Programming Interfaces (APIs) are not standardized, and often proprietary: only the BMS vendor can add functionality. Moreover, most of the devices that are used within contemporary buildings are not directly monitored or controlled via information systems because this would require the deployment

© Springer International Publishing Switzerland 2015
S. Gottwalt et al. (Eds.): EI 2015, LNCS 9424, pp. 59–68, 2015.
DOI: 10.1007/978-3-319-25876-8_6

of prohibitively expensive monitoring and control infrastructures. Still, increasing BMS coverage could lead to an entire ecosystem of innovative applications in the building sector, which could be further facilitated by opening up BMS APIs to third party application development. The final step toward smart buildings would then be to bring the human in the loop, i.e., enable individual but coordinated control of building services based on actual user feedback.

In this paper, we present the motivation, scope, and starting points of a new research project, NoFaRe, that will contribute to the realization of next generation BMS. NoFaRe will investigate innovative ways to realize device-level monitoring in buildings without adding sensors to every device, but infer the status of devices by applying machine learning methods to aggregated power signals, which has recently been referred to as Non-Intrusive Load Monitoring (NILM). It will also explore ways to integrate NILM capability into innovative BMS applications that can take full advantage of it.

The following Sect. 2 provides an overview of the state-of-the-art in BMS and NILM. Section 3 describes the scope of NoFaRe and how we plan to contribute to the technical landscape. Section 4 concludes our paper with a short summary and outlook.

2 Technical Landscape

2.1 Building Management Systems

Buildings can be viewed as complex cyber-physical systems consisting of many controllable elements, e.g., doors, windows, blinds, elevators, air conditioning units, lighting, fire protection, and various appliances. Although we are nowadays still used to controlling most of these elements manually, the degree of building automation is steadily increasing. BMS are software systems for monitoring and controlling the state of building elements. They rely on corresponding hardware, in particular sensors and actuators connected to a central server via a communication network. The visible part of a BMS typically includes a graphical user interfaces that allows building managers to remotely monitor relevant building functions and adjust controls whenever necessary.

Every building is unique, and so is its existing or potential management system. Apart from this unchangeable fact, however, there are more reasons for the abundance and heterogeneity of contemporary BMS. For instance, the building equipment for different functional areas, such as air conditioning and lighting, is often sourced from different vendors. Furthermore, the same company that provides a certain type of building equipment usually also develops and installs the corresponding BMS for monitoring and controlling the equipment. Thus, one often finds a separate BMS for different functional areas. When buildings get updated, even the equipment within the same functional area may become more heterogeneous, which can then results in several BMS per building function. Apart from certain communication protocol standards like BACnet, LonWorks, KNX or Modbus, which are typically used for the communication between a BMS and the sensors and actuators connected to it, little BMS standardization has happened so far. As a result, the different BMS within a building usually coexist as separate siloed systems. They are neither interoperable nor standardized [1].

The goal constraints of BMS are as diverse as the different functional areas they support. For instance, air conditioning systems are expected to sustain a comfortable indoor room climate, whereas lighting systems control the level of illumination in different areas of the building based on the need of its occupants. While these constraints should be met at all times, BMS should allow building managers to minimize cost, including maintenance and energy costs.

In both areas, i.e., the adherence to constraints and the maximization of building management goals, there still remains significant room for improvement. Given the energy and cost footprint of buildings, improvements in building control are a highly relevant research topic. In particular, the importance of information technology in buildings is quickly increasing. Current smart building research efforts proceed into several directions to improve the situation:

- Standardized BMS that could spur application innovation [1].
- Innovative control methods, in particular for indoor climate [2].
- Personalized control of building services [3].
- Sensor networks to increase measurement capability in buildings [4].

2.2   Non-Intrusive Load Monitoring

NILM was introduced by Hart and Schweppe in the late 1980's [5]. Their approach is based on continuously observing changes in the real and reactive power consumption measured at a single point in a circuit and detecting appliance on/off switching based on unique load signatures. This was sufficient to identify the state of small residential loads with a limited number of states at accuracies up to 85 % [5]. Hart's seminal work has spawned several follow-up studies investigating the feasibility of NILM in different settings using various methods, including work on more complex loads found in the commercial and industrial context [6]. Recently, several NILM frameworks featuring test data sets have been published [7,8]. Different NILM approaches can be classified according to the type of sensors and data granularity [9]. Table 1 contains pointers to recent NILM studies, including their data characteristics.

2.3   Research Gaps

In summary, both BMS and NILM are research areas that are attracting increasing attention in the computer science community. Furthermore, we believe that both topics are interrelated: NILM could allow for low-cost device-level monitoring, which could turn out as key enabler of next generation BMS. We believe that this connection has so far not been sufficiently appreciated. Rather, NILM research has so far focused on accurate disaggregation, but not deployment cost, usability, or concrete applications to building management. It is thus particularly interesting in our opinion to explore which innovative BMS applications could be enabled with NILM, which NILM algorithms are best suited for which

Table 1. Overview of Household Electricity Datasets

Dataset	Year	Project / University	Duration	Households	Aggregate Sampling
REDD [10]	2011	MIT	3–19 days	6	1 s & 15 kHz
BLUED [11]	2012	CMU	8 days	1	12 kHz
Smart* [12]	2012	UMass	3 months	3	1 s
Tracebase [13]	2012	Darmstadt	N/A	15	N/A[a]
IHEPCDS [14]	2012	University of California, Irvine	4 years	1	1 min
Sample [15]	2013	Pecan Street	7 days	10	1 min
OCTES [16]	2013	EU	4–13 months	33	7 s
HES [17]	2013	DECC, DEFRA	1/12 months	251	2 min
AMPds [18]	2013	Simon Fraser U	1 year	1	1 min
iAWE [19]	2013	IIIT Delhi	73 days	1	1 s
BERDS [20]	2013	University of California, Berkeley	7 days	1	20 s
UK-DALE [21]	2014	Imperial College	3–17 months	4	1–6 s & 16 kHz
GREEND [22]	2014	AAU Klagenfurt	1 year	9	1 s
ECO [8]	2014	ETH	8 months	6	1 s

[a] Available at device level.

application, and how actual NILM processes and system architectures could look like. In our opinion, more prototype-based research is needed to answer these questions and reveal practical challenges.

3   NoFaRe Project Scope

The NoFaRe project will address the research gaps mentioned above by developing a new NILM system including a self-designed smart meter box, the *NoFaRe Box*. The *NoFaRe Box* will be composed of a single-board computer equipped with a LAN/WiFi interface, a memory card, and one or multiple 16bit A/D converters for conducting circuit measurements at high frequencies of up to 44.1 kHz. Our NILM system will allow for device detection, type classification, state inference, and power disaggregation in an industrial, commercial, or private building environment. It will include a web-based management frontend that will allow for carrying out all necessary configuration, training and appliance registration tasks. Furthermore, we plan to design innovative BMS applications that leverage the NILM capabilities provided by the NoFaRe system. The targeted BMS features include the following:

Energy Saving. NILM capability can help to detect unnecessary energy consumption, e.g., by correlation of device-level consumption or benchmarking. It can also help to accurately identify inefficient devices and provide feedback to the responsible occupant or manager.

Maintenance Support. NILM can help to detect anomalous device behavior, which could be an indication of required maintenance. This feature can help to save manual inspection cost and increase service levels.

Energy Accounting. Device-level energy monitoring can pave the way toward energy billing according to actual consumption, possibly down to the level of individuals.

Safety. The ability to monitor single devices can enable new safety applications, ranging from the detection of potentially dangerous appliance states to detecting unauthorized access to buildings or machinery.

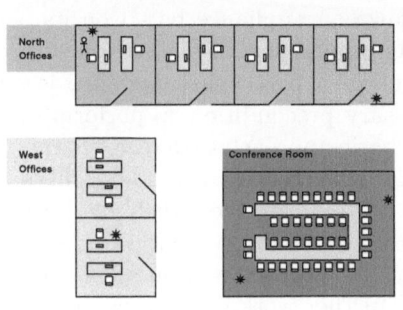

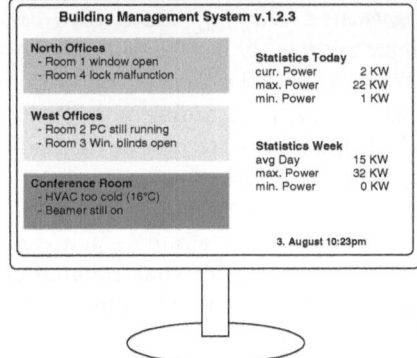

(a) Floor plan. NILM events shown as stars.

(b) BMS frontend. Event classification with localization.

Fig. 1. Outlook to future BMS using NILM techniques

Figure 1(a) and (b) show how NILM capability combined with floor plans can yield valuable information without the requirement to install expensive additional sensors.

3.1 Specific NILM Challenges

Based on the state-of-the-art of NILM, we will develop our own methods and corresponding architectures. One of our starting points is pattern recognition using high frequency current and voltage signals. We will develop methods that are able to determine the state of devices by detecting their characteristic influence on the current and voltage measured by the corresponding *NoFaRe Box*. In the following, we outline several challenges that we expect during the project:

Data Acquisition. Many highly useful appliance characteristics can only be observed at very short time scales. A good example of such a characteristic is the startup transient and inrush current, a very short but significant increase of the current as it energizes a devices. To measure such short time characteristics, it is necessary to obtain high frequency measurements. Sampling current at 1 Hz, as it is done in many NILM studies, is definitely insufficient for this purpose. Since we plan to make use of existing computer sound cards, the goal is a sampling frequency of up to 44.1 kHz.

Feature Extraction. To distinguish appliance classes (e.g., distinguishing hair dryers from television sets), discriminating features are necessary, which will be stored in the so-called feature space and used for the NILM tasks described in the following. A major challenge consists in finding features that are good representatives for one class but also offer a good discrimination against other classes. The extraction process of these characteristics is usually called *feature extraction* and uses all discriminating characteristics of the appliances.

Appliance On/Off Detection. To recognize an appliance based on its electricity signals, it is necessary to know if the appliance is switched on. On/off detection, i.e., the task of determining whether a particular appliance is currently in use (consuming power), is a necessary precondition to perform other NILM tasks, in particular disaggregation. A basic approach to detect whether an appliance is switched on is edge detection [5], which checks the power curve for step-like changes. A simple power threshold can indicate the inrush, as shown in Fig. 2. The method assumes that each device is consuming a measurable amount of power. We expect that reliable detection of multi-state appliances, such as dishwashers or dimmable lights, will require further work in the area of reliable edge detection.

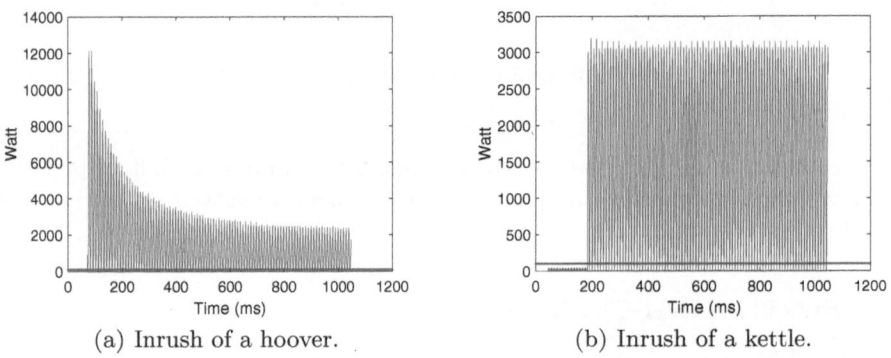

(a) Inrush of a hoover.   (b) Inrush of a kettle.

Fig. 2. Inrush comparison between a hoover and a kettle. The different characteristics are clearly visible. The *switch on* threshold is shown as red line (Colour figure online).

Appliance Identification. In building management, a number of appliances is usually known to be present. Appliance identification then refers to selecting those appliances from an existing list that are currently consuming power. Several studies have shown that this task can be carried out with relatively high accuracy [22].

Appliance Classification. Classification means assigning an appliance characteristic to a typical appliance class, e.g., washing machines, television sets, hairdryers etc. This task requires a set of typical characteristics for each class, which forms a distinguishable subspace in the feature space. A prerequisite for appliance classification is a well developed appliance taxonomy, which can be constructed manually, or automatically using clustering techniques. The work of [23] presents an approach to classify appliances using a clustering method based on their voltage-current (V-I) trajectory. Appliance classification becomes relevant if appliance identification fails because the observed characteristics do not match any registered appliances.

Disaggregation. The task of determining the individual consumption of several devices consuming power concurrently based on aggregated signals is referred to as disaggregation. A basic disaggregation approach is to first measure the amount of power that an individual device consumes over time and combine this information with edge detection. Many more sophisticated disaggregation methods have been evaluated, e.g., Factorial Hidden Markov Model (FHMM) [10]. Due to its complexity and potential impact, disaggregation can certainly be considered as one of the most demanding NILM tasks at the moment.

Privacy Preservation. NILM raises many privacy concerns because it potentially allows for tracking human behavior. In fact, NILM is not limited to determining what happens in a household, e.g., sleeping, being away, cooking, watching television. It even allows for going further, e.g., identifying the television program currently being watched [24,25]. An important challenge of NILM is therefore to preserve privacy according to legal requirements, i.e., certain information must not be used without the consent of the concerned individuals and must be effectively protected from other uses. In NoFaRe, we will therefore pay close attention to the secure processing and transmission of potentially revealing data.

3.2   A Tentative Appliance Identification Algorithm

Figure 3 shows the flow chart of a tentative appliance detection algorithm. We present it here to show how several NILM tasks can be combined into a continuous process that could later run on the *NoFaRe box*.

The algorithm as two parallel loops. The first loop continuously acquires the current and voltage from the A/D converters and observes their characteristics. If this observation indicates that a device has just been switched on, feature extraction starts in a new thread. Based on the extracted features, the algorithm then tries to identify the added appliance based on the devices previously registered by the user. If this appliance identification is successful (based on a corresponding confidence threshold), the identified user-registered appliance is added to the list of *currently running* appliances. If the appliance identification is not confident, an appliance classification step will be triggered to assign the new device to a device type present in the local *appliance type* data base. Since multi-state appliances, like washing machines, can be certainly recognized after

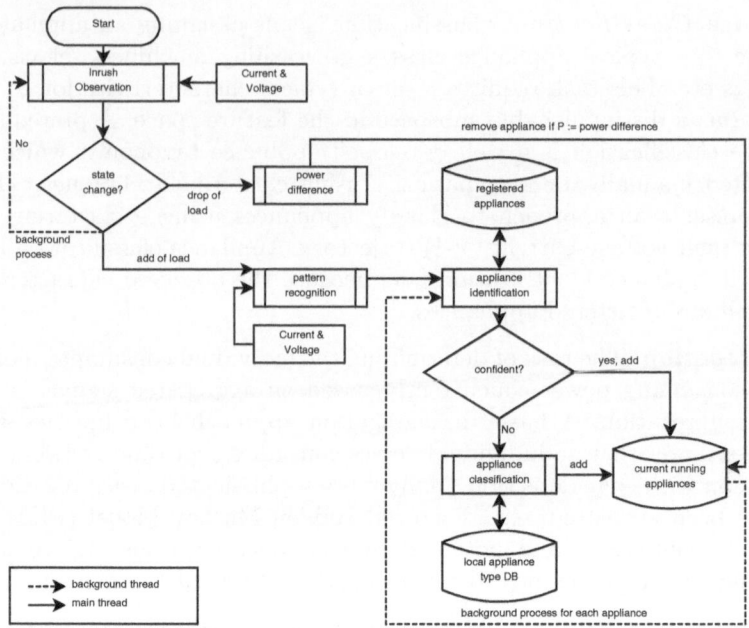

Fig. 3. Flow chart visualization of the *NoFaRe* appliance detection algorithm.

some time has passed, it is necessary to continuously repeat the classification step in a second, parallel loop.

3.3   Preliminary Experiments and Results

First experiments with the proposed algorithm for appliance detection, based on their startup transient characteristics, yielded promising results. The experimental setup is based on *House 1* of the UK-DALE dataset [21]. We have implemented an automatic inrush recognition based on short time load changes to find the actual startup of the appliances. The rough positions are manually retrieved from the 6 s data set [26] and a 15 s window is extracted from the 16 kHz dataset [27] for further automated analysis. After finding the exact startup time and extracting the first 500 ms, we extract 5 discriminating features of 12 different appliance types from around 10,000 samples. Based on 10-fold cross validation, the classification accuracy lies consistently above 85 % with a 5-nearest neighbor classifier.

4   Conclusion

In this paper, we have introduced the idea of integrating NILM technology into the BMS landscape. NILM can enable new BMS that take advantage of high resolution monitoring capability at lower cost compared to dedicated sensor networks. In the NoFaRe project, we intend to evaluate the presented concept by developing a corresponding system architecture. A core element of our prototypical system will be the NoFaRe box, a low-cost single-board computer that will measure electricity signals at high frequency and perform continuous device detection. Since the NoFaRe project is still in an early stage, we have only presented preliminary evaluation results. However, we have placed NoFaRe within the current technical landscape, detailed its expected contribution, and described several challenges which we expect to face during the project.

References

1. Dawson-Haggerty, S. et al.: BOSS: Building Operating System Services. In: Proceedings of the 10th USENIX Conference on Networked Systems Design and Implementation. ACM, pp. 443–458 (2013)
2. Aswani, A. et al.: Reducing transient and steady state electricity consumption in HVAC using learning-based model-predictive control. In: Proceedings of the IEEE 100.1 (2012), pp. 240–253 (2011). doi:10.1109/jproc.2011.2161242
3. Krioukov, A. et al.: A living laboratory study in personalized automated lighting controls. In: Proceedings of the Third ACM Workshop on Embedded Sensing Systems for Energy-Efficiency in Buildings. ACM, pp. 1–6 (2011). doi:10.1145/2434020.2434022
4. Jiang, X. et al.: Experiences with a high-fidelity wireless building energy auditing network. In: Proceedings of the 7th ACM Conference on Embedded Networked Sensor Systems. ACM, pp. 113–126 (2009). doi:10.1145/1644038.1644050
5. Hart, G.W.: Nonintrusive appliance load monitoring. In: Proceedings of the IEEE 80.12, pp. 1870–1891 (1992). doi:10.1109/5.192069
6. Lee, K.D., et al.: Estimation of variable-speed-drive power consumption from harmonic content. IEEE Trans. Energy Convers. **20**(3), 566–574 (2005). doi:10.1109/tec.2005.852963
7. Batra, N. et al.: NILMTK: an open source toolkit for non-intrusive load monitoring. In: Proceedings of the 5th International Conference on Future Energy Systems, pp. 265–276 (2014).doi:10.1145/2602044.2602051
8. Beckel, C. et al.: The ECO data set and the performance of non- intrusive load monitoring algorithms. In: Proceedings of the 1st ACM Conference on Embedded Systems for Energy-Efficient Buildings, pp. 80–89 (2014). doi:10.1145/2674061.2674064
9. De Paola, A. et al.: Intelligent management systems for energy efficiency in buildings: a survey. In: ACM Computing Surveys (CSUR) 47.1, p. 13 (2014). doi:10.1145/2611779
10. Kolter, J.Z., Johnson, M.J.: REDD: a public data set for energy disaggregation research. In: Workshop on Data Mining Applications in Sustainability (SIGKDD), vol. 25, pp. 59–62. Citeseer, San Diego (2011)
11. Anderson, K. et al.: BLUED: a fully labeled public dataset for event- based non-intrusive load monitoring research. In: Proceedings of the 2nd KDD Workshop on Data Mining Applications in Sustainability (SustKDD), pp. 1–5 (2012)
12. Barker, S. et al.: Smart*: an open data set and tools for enabling research in sustainable homes. In: SustKDD (2012)
13. Reinhardt, A. et al.: Electric appliance classification based on distributed high resolution current sensing. In: 37th Annual IEEE Conference on Local Computer Networks, Workshop Proceedings, pp. 999–1005 (2012). doi:10.1109/lcnw.2012.6424093

14. Lichman, M.: UCI machine learning repository: individual household electric power consumption data set (2013). http://archive.ics.uci.edu/ml
15. Pecan Street Inc., Sample Data Set (2013).http://pecanstreet.org/projects/ consortium (online accessed 02 June 2015)
16. European Union. Opportunities for Community Groups Through Energy Storage (OCTES) (2013). http://octes.oamk.fi/final (online accessed 02 June 2015)
17. Zimmermann, J.-P. et al.: Household electricity survey: a study of domestic electrical product usage. In: Intertek Testing & Certification Ltd (2012)
18. Makonin, S. et al.: AMPds: a public dataset for load disaggregation and eco-feedback research. In: Electrical Power & Energy Confer- ence (EPEC). IEEE, pp. 1–6 (2013).doi:10.1109/epec.2013.6802949
19. Batra, N. et al.: It's different: insights into home energy consumption in India. In: Proceedings of the 5th ACM Workshop on Embedded Systems for Energy-Efficient Buildings, pp. 1–8. ACM (2013). doi:10.1145/2528282.2528293
20. Maasoumy, M. et al.: Berds-Berkeley energy disaggregation data set. In: Proceedings of the Workshop on Big Learning at the Conference on Neural Information Processing Systems (2014)
21. Kelly, J., Knottenbelt, W.: The UK-DALE dataset, domestic appliance-level electricity demand and whole-house demand from five UK homes. In: Scientific Data (2015). doi: 10.1038/sdata.2015.7. http://www.doc.ic.ac.uk/dk3810/data/
22. Monacchi, A. et al.: GREEND: an energy consumption dataset of households in Italy and Austria. In: CoRR abs/1405.3100 (2014). doi:10.1109/smartgridcomm. 2014.7007698
23. Lam, H.Y., Fung, G.S.K., Lee, W.K.: A novel method to construct taxonomy electrical appliances based on load signatures. IEEE Trans. Consum. Electron. **53**(2), 653–660 (2007). doi:10.1109/tce.2007.381742
24. Greveler, U., Justus, B., Löhr, D.: Hintergrund und experimentelle Ergebnisse zum Thema Smart Meter und Datenschutz. In: Fachhochschule Munster-University of Applied Sciences Technical Paper (2011)
25. Enev, M. et al.: Televisions, video privacy, and powerline electromagnetic interference. In: Proceedings of the 18th ACM Conference on Computer and Communications Security, CCS 2011, pp. 537–550. ACM, Chicago (2011). ISBN: 978-1-4503-0948-6, doi:10.1145/2046707.2046770
26. UK-DALE: disaggregated (6s) appliance power and aggregated (1s) whole house power (2015). doi:10.5286/UKERC.EDC.000001
27. UK-DALE: high speed (16 kHz) aggregated whole house current/voltage (2015). doi:10.5286/UKERC.EDC.000002

The Effect of Data Granularity on Load Data Compression

Andreas Unterweger[1](✉), Dominik Engel[1],
and Martin Ringwelski[2]

[1] Josef Ressel Center for User-Centric Smart Grid Privacy, Security and Control,
Salzburg University of Applied Sciences, Urstein Süd 1, 5412 Puch/Salzburg, Austria
{Andreas.Unterweger,Dominik.Engel}@en-trust.at
[2] Institut für Telematik, Technische Universität Hamburg-Harburg,
Am Schwarzenberg-Campus 1, 21073 Hamburg, Germany

Abstract. A vast volume of data is generated through smart metering. Suitable compression mechanisms for this kind of data are highly desirable to better utilize low-bandwidth links and to save costs and energy. To date, the important factor of data resolution has been neglected in the compression of smart meter data. In this paper, we review and evaluate compression methods for smart metering in the context of different resolutions. We show that state-of-the-art compression methods are well suited for high resolution, but not for low resolution data. Furthermore, we elaborate on the compression performance differences between appliance-level and household-level load data. We conclude that the latter are practically incompressible at most resolutions.

1 Introduction

In smart grids, the volume of data to be processed, transmitted and stored is considerable. In the distribution grid, smart meters are a source of high data volume. Depending on the use case and regulatory restrictions, different measurements are collected by a smart meter in different granularities, typically in measurement intervals of 60 s up to 15 min (cf. Table 10 in [7]). Smart meters are also capable of collecting measurements related to power quality. All measurements can technically be done in smaller intervals (i.e., seconds).

It is evident that compressing the data generated in smart metering is highly desirable. Smart meters are typically connected via low-bandwidth links, such as PLC. Through compression, the bandwidth of these links can be utilized more efficiently. The increase in efficiency, of course, depends on the measurement interval and will increase with smaller intervals. Furthermore, transmitting data in compressed form is more energy-efficient than transmitting data in uncompressed form – given that an appropriately light-weight compression scheme is used, the power needed for compression is significantly lower than the power needed for transmission. Finally, at the receiving end, where the data needs to be stored, compression can help to save costs.

© Springer International Publishing Switzerland 2015
S. Gottwalt et al. (Eds.): EI 2015, LNCS 9424, pp. 69–80, 2015.
DOI: 10.1007/978-3-319-25876-8_7

It comes at no surprise that a number of proposals have been made for the compression of smart metering data. However, none of these contribution explicitly addresses the issue of resolution and its impact on compression performance. While there are many benefits to compression, it has to be evaluated how well the raw data is suited for compression at different measurement intervals, i.e., varying data granularity. An overview of standard compression methods applied to smart metering data is given by Ringwelski et al. [11]. Furthermore, the authors propose their own method. Unterweger and Engel [13] propose a compression method that allows resumability. Two contributions that implicitly address resolution, because they both employ the wavelet transform for compression are Ning et al. [10] and Khan et al. [8]. However, neither takes the impact of resolution of the *source* data into account.

In general, there is little research that addresses the resolution of smart metering data, mostly in the area of smart meter privacy. Eibl and Engel [5] give an account on the influence of data granularity on privacy in smart metering. Approaches for privacy-preserving smart metering are presented by Efthymious and Kalogridis [4] and Engel [6]. Sankar et al. [12] introduce an information-theoretic framework for smart meter privacy, which implicitly addresses data resolution as part of the proposed privacy measure.

In this paper, we evaluate the compression algorithms proposed by Ringwelski et al. [11] and Unterweger and Engel [13] in the context of source data resolution. This is an important perspective, as different use cases in the smart grid will require different measuring intervals and therefore different resolutions of load data. An appraisal on how this resolution impacts compression performance gives an important guideline on what amount of data needs to be transmitted for the individual smart metering use cases.

This paper is structured as follows: In Sect. 2, we describe the compression algorithms that we evaluate in Sect. 3. Section 4 concludes.

2   Compression Algorithms

Several algorithms for compressing load data have been studied in the literature. We focus on those algorithms which have been specifically designed for load data in the context of smart metering, where resources are typically sparse, i.e., execution time and memory consumption have to be minimized.

For reference, we use two standardized encodings for load data which do not compress the data. For our measurements in Sect. 3, we use two tailored compression algorithms. All four approaches are described in the following sections. Although some encodings specify the use of units (e.g., watts), we focus on the value encoding only. Unit signaling can be amended if necessary, but is out of scope of this work.

2.1   Reference Algorithms

Two standards for transmitting load data are commonly used: IEC 62056-21 [3] and IEC 61334-6 [2], also referred to as A-XDR. Both specify value encodings which do not perform any compression whatsoever, minimizing computational complexity. In the following, we describe both algorithms briefly since we use them for reference measurements.

IEC 62056-21. Values are encoded in their base 10 representation with a decimal point and encoded as ASCII [1] bytes. The value *123.45*, for example, is encoded as 00110001 00110010 00110011 00101110 00110100 00110101, requiring six bytes – five digits and the decimal point.

Since the length of each encoded value depends on its magnitude, an additional delimiter between subsequent values is required so that they can be separated during decoding. Without additional signaling information, an underscore (ASCII character 137), for example, can be used as a delimiter. This way, the values *123.45* and *123.56*, for example, are concatenated to *123.45_123.56* before encoding, requiring a total of $6 + 1 + 6 = 13$ bytes.

A-XDR. Unsigned integer values are encoded in their base 2 representation with a fixed length, e.g., 16 bits. The value *12345*, for example, is encoded as 00110000 00111001, requiring 2 bytes. Although floating-point values are not supported, multiplying the floating-point value by 10^n, where n is the number of decimal places after the decimal point, yields an integer value which can be encoded using A-XDR.

Since the number of decimal places does typically not change within a load data time series, no additional signaling for n is required. However, the number of bits required for representation may have to be increased to accommodate for the increased value range due to the multiplication by 10^n. For example, encoding the value *123.45* (as *12345*, see above) requires at least 14 bits, as opposed to the value *123*, which only requires 7 bits.

As stated above, A-XDR coding uses a fixed bit length for representing values. Thus, all values can be decoded without the need for any additional delimiters as opposed to the IEC 62056-21 value coding described above.

2.2 DEGA Coding

Unterweger and Engel [13] have proposed a compression algorithm for load data which exploits the data characteristics of load profile data. Their encoding algorithm, which we refer to as DEGA (Differential Exponential Golomb and Arithmetic) coding due to its main elements, is illustrated in Fig. 1 and consists of five steps (labeled A-E).

First, the floating-point input values are normalized (A) to make them integer, as explained for A-XDR in Sect. 2.1. Second, the differences between consecutive values are calculated (B), since they are typically smaller than the values themselves. Third, the differences are encoded as Signed Exponential Golomb code words of order zero (C) for variable-length coding. Fourth, the code words are concatenated (D) and finally compressed using an adaptive binary arithmetic coder (E).

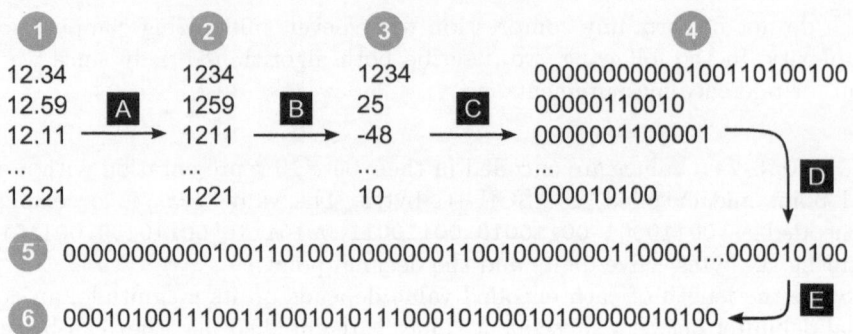

Fig. 1. Overview of DEGA coding [13]: Input values (1) are normalized and their differences (3) are represented as Signed Exponential-Golomb code words (4) which are concatenated (5) and arithmetically coded.

During processing, the code word concatenation step (D) is usually implicitly contained in the code word generation step (C). A detailed explanation of each step as well as a description of the decoding process can be found in [13].

2.3   LZMH Coding

Ringwelski et al. [11] have proposed a compression algorithm for load data with low memory requirements. The algorithm is referred to as Lempel Ziv Markov Chain Huffman (LZMH) coding and combines ideas of the Lempel Ziv Markov Chain Algorithm (LZMA) and a variant of Adaptive Trimmed Huffman (AHT) coding as described below and illustrated in Fig. 2. It is designed to process ASCII-coded IEC 62056-21 data as described in Sect. 2.1 as input.

If at least three of the following characters are found in the history of the last m characters, a reference to it is coded (LZMA-like), consisting of a byte offset and the length, using an optimal prefix code. Conversely, when no sufficient reference is found, it is encoded as a Huffman code word (AHT-like). This code word originates from an adaptive Huffman tree which represents the symbol probabilities that are updated for each encoded character.

To keep memory requirements low, a history buffer of $m = 128$ characters is used and the size of the Huffman tree is limited to the size of the input alphabet which may be reduced to the ten decimal digits and the decimal point. A more detailed description of the algorithm can be found in [11].

3   Evaluation

We analyze the compression performance and execution times of A-XDR, DEGA and LZMH coding for IEC 62056-21 input data. The used data sets are described in detail in Sect. 3.1. As opposed to prior work, we study the effect of different data granularity on the results.

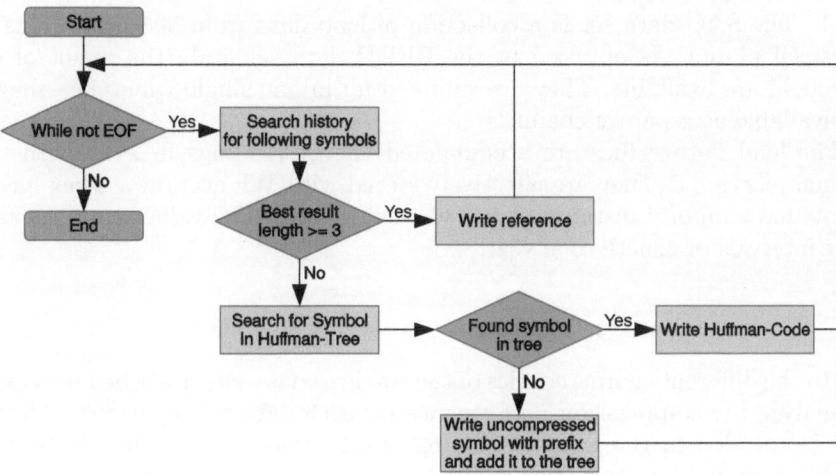

Fig. 2. Overview of LZMH coding: Each input symbol is either encoded as a reference to an already processed symbol or as a Huffman code word based on its probability.

We evaluate different data granularity levels by summing up c consecutive input data values with inter-value temporal distance t, for example, 5 min (300 s) granularity for $t = 3$ (seconds) with $c = 100$. We use the same granularity levels as Eibl and Engel [5], i.e., 3 s, 9 s, 30 s, 1 min, 5 min, 15 min and 1 h, if available.

To achieve comparable results, we have reimplemented the A-XDR and DEGA coding algorithms in the C programming language. LZMH is already implemented in C and has only been modified slightly so that it uses the same input/output functions. These changes do not affect its compression performance.

3.1 Load Data Sets

We use two load data sets for our evaluation: the low-frequency MIT REDD data set [9] and a data set from a local energy provider, referred to as the SAG data set henceforth. Both data sets are described briefly below.

REDD. The low-frequency MIT REDD data set is a collection of load data from between 11 and 26 channels of 6 different houses. In total, there are 116 channels. Each channel containing load data is available separately.

The load data values are average apparent power readings in Watts with two decimal places, i.e., they are effectively stored with an accuracy of one hundredth of a Watt. They have an inter-value temporal distance of $t = 3$ (seconds) for all channels but the mains, which have $t = 1$. The values cover measurement intervals of between 2.7 and 25.8 days.

SAG. The SAG data set is a collection of load data from 508 households and industrial plants. As opposed to the REDD data set, only the mains of each household are available. They are summed up in one single value, i.e., they are not available as separate channels.

The load data values are accumulated energy readings in kWh with three decimal places, i.e., they are effectively stored with Wh accuracy. They have an inter-value temporal distance of $t = 300$, i.e., 15 min. The values cover measurement intervals of exactly one year.

3.2   Compression Performance

Due to the different characteristics of the two load data sets described in Sect. 3.1, we analyze the compression performance for each data set separately. All input data is encoded in the form of IEC 62056-21 values as described in Sect. 2.1, which we use as reference. The results are described in the following sections.

REDD Data Set. Figure 3 shows an overview of the compression performance of the A-XDR, DEGA and LZMH algorithms for the REDD data set. Each channel is compressed separately and its compressed size is expressed relative to the input data size as a ratio. A compression ratio of 5, for example, means that the compressed data requires only 20 % of the size of IEC 62056-21 value encoding.

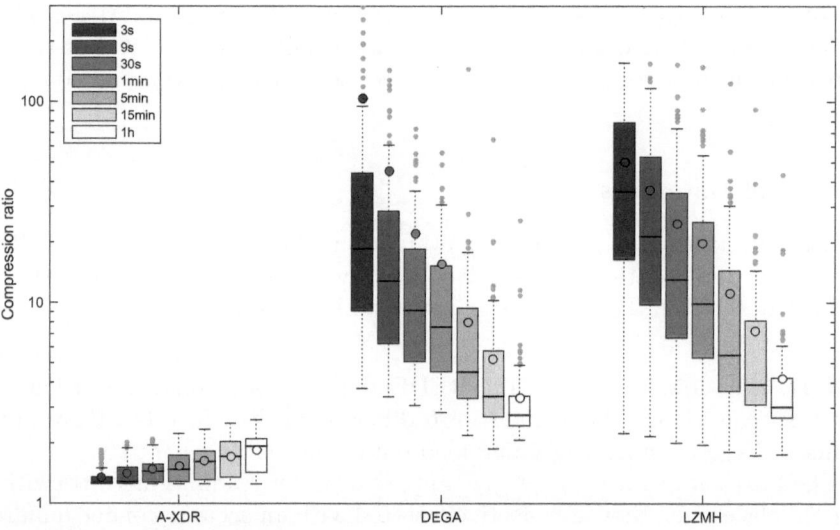

Fig. 3. Compression performance of different algorithms compared to IEC 62056-21 value encoding for the REDD load data set at different data granularity levels.

The compression ratio distribution for all channels is depicted as a box plot with added mean compression ratios (filled circles with black borders) and outliers (gray circles without borders). The y axis is logarithmic and capped at 300. Thus, four outliers representing all-zero valued channels are not depicted.

Obviously, DEGA and LZMH exhibit significantly better compression performance than A-XDR, which does not compress by design. Still, it achieves compression ratios greater than 1 compared to IEC 62056-21 value encoding. This is due to the fact that all input values are at least five bytes long (one decimal digit before the decimal point, two thereafter and one delimiter), but typically longer, whereas A-XDR values are always four bytes in size.

In general, LZMH outperforms DEGA at all granularity levels, where the performance difference increases with data granularity. At the finest granularity level (3 s, dark gray boxes), DEGA and LZMH achieve compression ratios of 18.59 and 35.48, respectively. They drop to 2.77 and 3.02, respectively, at the coarsest granularity level (1 h, white boxes).

Compared to A-XDR coding with a median compression ratio of 1.94 at this granularity level, it becomes clear that both, DEGA and LZMH, are practically ineffective at compressing load data with high (1 h) inter-value temporal distances. A-XDR is expected to outperform both compression algorithms at even coarser granularity levels, e.g., at inter-value temporal distances of 24 h.

In general, increased inter-value temporal distances yield larger input values, i.e., they have more decimal digits and therefore yield longer IEC 62056-21 values. Since A-XDR values are of constant size, their compression ratio increases relatively at coarser granularity levels, whereas DEGA and LZMH coding become less efficient in terms of compression performance. This is mainly due to the increased input entropy.

Coarser data granularity impacts compression performance due to the summing of values. Thus, the mains (channels 1 and 2) of all houses from the REDD data set deserve special attention. They, too, are effectively sums of multiple other channels and therefore likely to behave differently than the other channels. Figure 4 shows the compression performance of only the mains.

As expected, the compression performance of DEGA and LZMH coding for the mains is significantly poorer than the respective performance for all channels depicted in Fig. 3. Although the best median compression ratio for fine-grain data (3 s, dark gray in Fig. 4) is 4.90, double-digit compression performance is not achievable for the mains.

Interestingly, when compressing only the mains, DEGA outperforms LZMH at all granularity levels. The reverse is true when looking at the compression performance of all channels in Fig. 3. Still, at medium granularity levels (1 min, medium gray in Fig. 4), compression becomes ineffective when compared to uncompressed A-XDR coding.

Even more surprisingly, at coarser granularity levels (15 min, light gray), A-XDR actually outperforms DEGA coding with a median compression ratio of

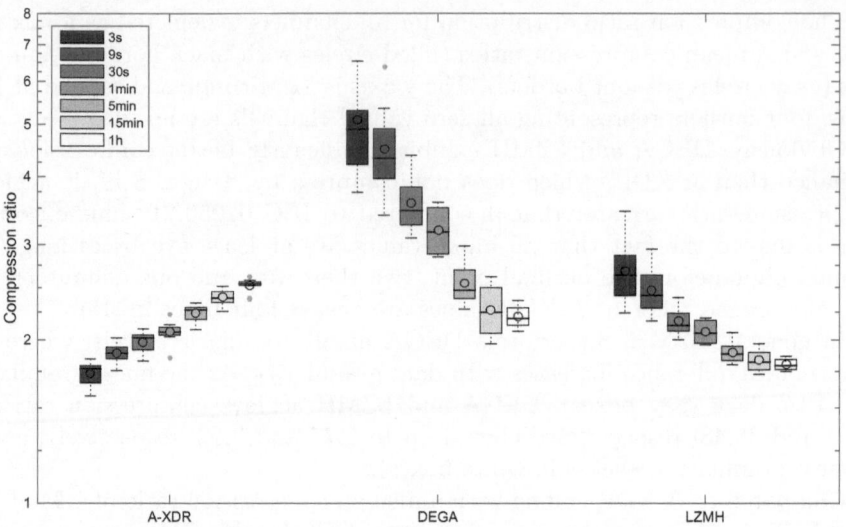

Fig. 4. Compression performance of different algorithms compared to IEC 62056-21 value encoding for only the mains from the REDD load data set.

2.39 vs. 2.25 despite the fact that A-XDR does not compress by design. This means that the mains are effectively incompressible at this resolution.

SAG Data Set. Figure 5 shows the compression results of the A-XDR, DEGA and LZMH algorithms for the SAG load data set. Since the latter only has 15-minute resolution, finer granularity levels cannot be evaluated. The visualization is identical to the one in Fig. 3 for the REDD data set.

Since the SAG data set only contains measurements from the mains and not from individual channels, the results are similar to the results of the mains from the REDD data set illustrated in Fig. 4. Again, DEGA outperforms LZMH coding, but the compression ratios of both are higher when compared to A-XDR, i.e., the data can be compressed to some extent, even at a temporal inter-value distance of 1 h.

The main reason for this, considering that the mains from the REDD data set are incompressible as explained above, is the different accuracy of the data. While the REDD mains data has an accuracy of one hundredth of a Watt, the SAG data has an accuracy of only one Watt-hour. This significantly reduces the entropy since the highly volatile least significant digits are missing.

Apart from the lower accuracy, the value range is also reduced, i.e., the kWh readings (SAG) are significantly smaller than Watt readings (REDD). This also explains the high number of outliers (gray circles without borders) in Fig. 5 for DEGA and LZMH coding: Households with a lower power consumption yield smaller values which can be more compressed more easily.

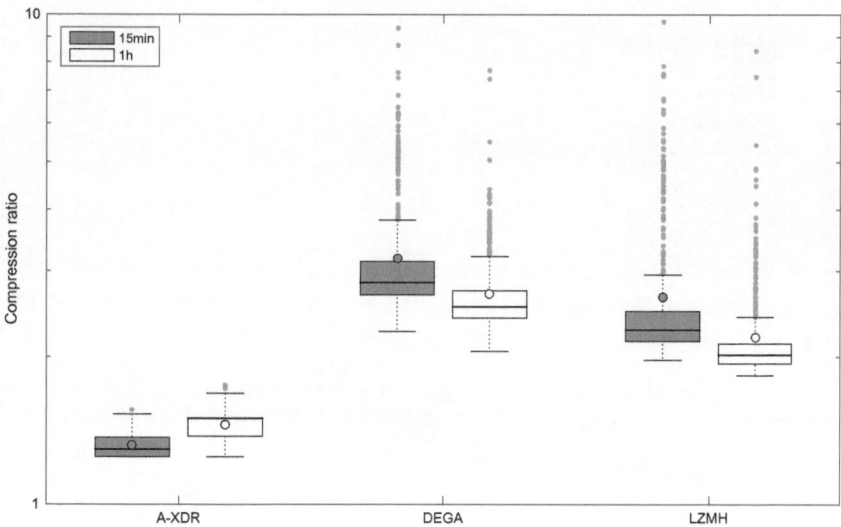

Fig. 5. Compression performance of different algorithms compared to IEC 62056-21 value encoding for the SAG load data set at different data granularity levels.

3.3   Execution Time

The DEGA and LZMH compression algorithms reduce the data rate in a number of cases as described above. However, they are computationally more complex than uncompressed data transmission. Thus, the additional code execution time has to be analyzed.

We measure the execution time similar to Unterweger and Engel [13]. Each channel (REDD data set) or household (SAG data set) is processed, as a whole, three times for cache warming and then five times for the actual time measurements. The five time results are averaged and divided by the number of data values in the processed channel/household to yield the average processing time per data value.

Again, the REDD and SAG data sets are evaluated separately due to their different data characteristics. A-XDR encoding is used as reference for uncompressed processing. All results have been obtained on a virtualized 64-bit *Ubuntu* 14.04 machine with *gcc* 4.8.2 running on an Intel Xeon W3503 CPU.

REDD Data Set. Figure 6 shows an overview of the execution time per data value required by the A-XDR, DEGA and LZMH algorithms for the REDD data set. Despite the powerful CPU used for benchmarking, the processing time is in the microsecond range, i.e., most likely in the 10- or 100-microsecond range on less powerful hardware, e.g., smart meters. This can be considered feasible.

LZMH coding is clearly faster than DEGA coding. Surprisingly, it is, in the majority of cases, even faster than uncompressed A-XDR coding. This can

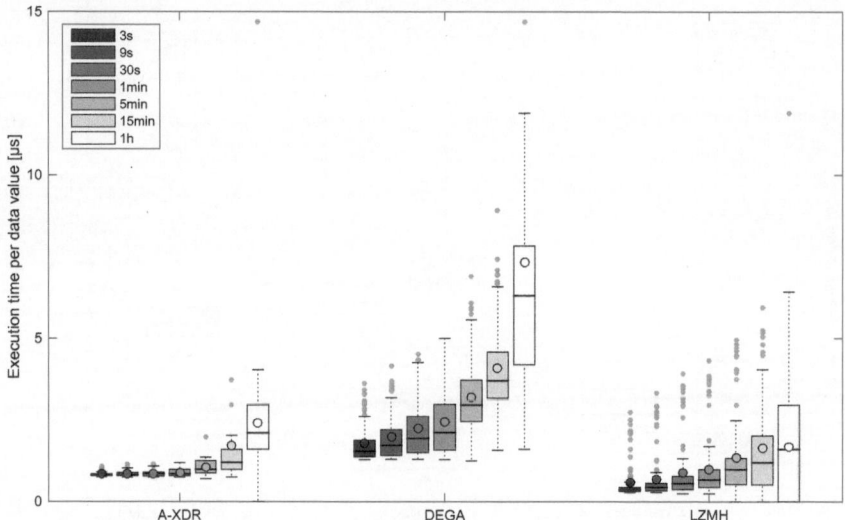

Fig. 6. Execution times per value of different algorithms when compressing channels of the REDD load data set at different data granularity levels.

be explained by the fact that both algorithms process data on a per-character basis, but A-XDR requires a conversion to floating point values which involves expensive floating point operations. These require about as much time as the whole compression step of LZMH coding, which is very compact.

Execution times increase at finer data granularity levels for all algorithms due to the relative increase in size of the input data. Since the (summed) values are larger in terms of magnitude, their IEC 62056-21 representations are longer. This explains the increased slopes of the median execution times for A-XDR and DEGA coding at coarser granularity levels. As LZMH coding does not convert the representation of the values, its slope is not affected by their magnitude, but by their redundancy, resulting in a smaller slope.

SAG Data Set. Figure 7 shows an overview of the execution time per data value required by the A-XDR, DEGA and LZMH algorithms for the SAG data set. The visualization is identical to the one in Fig. 6 for the REDD data set, with the exception of the data granularity range due to the 15-minute inter-value temporal distance of the original data.

The order of magnitude of the execution times is the same as for the REDD data set. However, the absolute values are lower for all algorithms due to the smaller (and therefore shorter) input values. Interestingly, also the differences between 15 min and 1 h granularity are significantly smaller for the SAG data set than for the REDD data set. Again, this is due to the range (and therefore the length) of the input values.

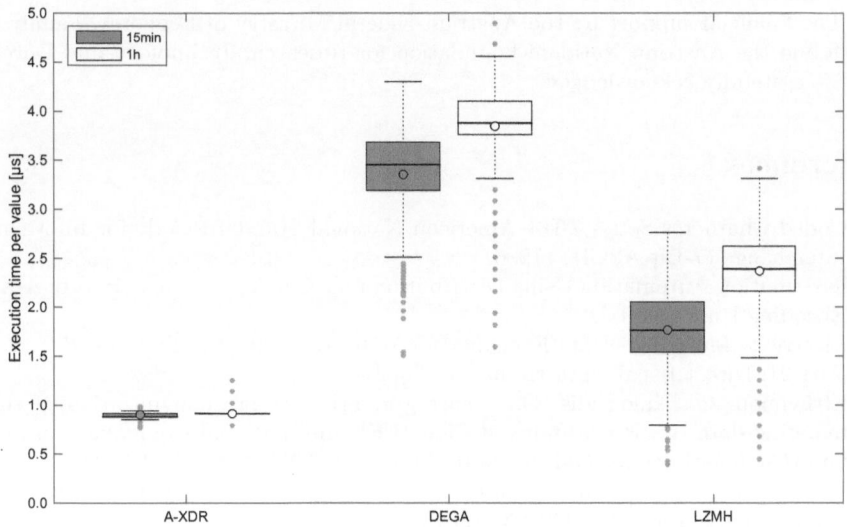

Fig. 7. Execution times per value of different algorithms when compressing households of the SAG load data set at different data granularity levels.

The slope between the 15 min and 1 h granularity levels for the SAG data set in Fig. 7 is comparable to the slope for 3 s to 5 min granularity levels for the REDD data set in Fig. 6. This shows that both, execution times and execution time differences are highly dependent on the input data length.

In addition, the differences in terms of execution time between DEGA and LZMH coding are smaller. This is due the lower compression efficiency of LZMH. This also explains why, in contrast to the execution times for the REDD data set (see Fig. 6), LZMH is slower than A-XDR coding for the SAG data set.

4   Conclusion

Load data from the evaluated data sets is compressible, but only at fine data granularity levels, e.g., 3 s intervals. At coarser granularity levels, compression becomes less effective or even futile, i.e., the reduction in data rate is practically insignificant compared to uncompressed encoding. This effect is stronger for the mains of a household than for per-room or per-device channels which have lower entropy and are therefore easier to compress. When compressing the tested load data sets, LZMH coding by Ringwelski et al. is recommended for the latter type of channels at fine data granularity levels. For coarser granularity levels as well as the mains, DEGA coding by Unterweger and Engel offers higher compression ratios at the cost of longer execution time.

Acknowledgements. The authors would like to thank Günther Eibl for his help in visualizing the compression ratio results. They would also like to thank their partner Salzburg AG for providing additional real-world load data.

The financial support by the Austrian Federal Ministry of Economy, Family and Youth and the Austrian National Foundation for Research, Technology and Development is gratefully acknowledged.

References

1. Coded Character Sets - 7-Bit American National Standard Code for Information Interchange (7-Bit ASCII) (1986)
2. Distribution Automation Using Distribution Line Carrier Systems - Part 6: A-XDR Encoding Rule (2000)
3. Electricity Metering - Data Exchange for Meter Reading, Tariff and Load Control - Part 21: Direct Local Data Exchange (2002)
4. Efthymiou, C., Kalogridis, G.: Smart grid privacy via anonymization of smart metering data. In: Proceedings of First IEEE International Conference on Smart Grid Communications, Gaithersburg, Maryland, USA, pp. 238–243 (2010)
5. Eibl, G., Engel, D.: Influence of data granularity on smart meter privacy. IEEE Trans. Smart Grid 6(2), 930–939 (2015)
6. Engel, D.: Wavelet-based load profile representation for smart meter privacy. In: Proceedings of IEEE PES Innovative Smart Grid Technologies (ISGT 2013), Washington, D.C., USA, pp. 1–6 (2013). http://dx.doi.org/10.1109/ISGT.2013.6497835
7. European Commission: Cost-benefit analyses & state of play of smart metering deployment in the EU-27. Technical report, European Commission Report (2014). http://eur-lex.europa.eu/legal-content/EN/TXT/PDF/?uri=CELEX:52014SC0189&from=EN
8. Khan, J., Bhuiyan, S., Murphy, G., Arline, M.: Embedded zerotree wavelet based data compression for smart grid. In: 2013 IEEE Industry Applications Society Annual Meeting, pp. 1–8 (2013)
9. Kolter, J., Johnson, M.J.: Redd: a public data set for energy disaggregation research. In: Workshop on Data Mining Applications in Sustainability (SIGKDD), pp. 1–6, August 2011
10. Ning, J., Wang, J., Gao, W., Liu, C.: A wavelet-based data compression technique for smart grid. IEEE Trans. Smart Grid 2(1), 212–218 (2011)
11. Ringwelski, M., Renner, C., Reinhardt, A., Weigel, A., Turau, V.: The Hitchhiker's guide to choosing the compression algorithm for your smart meter data. In: 2012 IEEE International Energy Conference and Exhibition (ENERGYCON), pp. 935–940, September 2012
12. Sankar, L., Rajagopalan, S.R., Mohajer, S., Poor, H.V.: Smart meter privacy: a theoretical framework. IEEE Trans. Smart Grid 4(2), 837–846 (2013)
13. Unterweger, A., Engel, D.: Resumable load data compression in smart grids. IEEE Trans. Smart Grid 6(2), 919–929 (2015). http://dx.doi.org/10.1109/TSG.2014.2364686

Research Lab Infrastructures

A Concept for the Control, Monitoring and Visualization Center in Energy Lab 2.0

Clemens Düpmeier, Karl-Uwe Stucky$^{(\boxtimes)}$, Ralf Mikut,
and Veit Hagenmeyer

Karlsruhe Institute of Technology (KIT), Karlsruhe, Germany
{clemens.duepmeier,uwe.stucky,ralf.mikut,
veit.hagenmeyer}@kit.edu

Abstract. Energy Lab 2.0 is designed as a large experimental test and simulation field for multi-scale and multi-mode energy system facilities at KIT. A Smart Energy System Simulation and Control Center (SEnSSiCC) is the core component in terms of information and communication technology. The present article introduces basic concepts for the Control, Monitoring and Visualization Center (CMVC) of SEnSSiCC. The CMVC bundles all communication channels and real facilities, simulation environments, and data repositories into an integrated research environment for planning, control, monitoring, analyzing and visualization of smart grids and their components, and furthermore for evaluating future concepts for smart grid utility operation. Special emphasis is placed on the distributed computing operating system environment setup for the CMVC, the intended use of Big Data technologies, the polyglot approach for data management and analysis, and first concepts for implementing a hybrid agent based simulation environment. Also, the usage of web technologies and microservices are considered as key aspects of the overall architecture.

Keywords: Energy lab · Energy system · Microgrid · Simulation · Visualization · Web technology

1 Introduction

The future energy system will be characterized by efficient conversion of mostly regenerative primary energy to power, heat and fuels as secondary energy carriers. As an example, the German government has launched an energy transition plan with a nuclear phase-out as soon as 2022 and aiming for 80 % renewable primary energy in 2050 [21]. Challenges derive from spatial and temporal fluctuations in power generation by regenerative sources that meet the already existing fluctuations in consumer demand and also from various technological aspects [3]. This leads to enforcing of new business models in energy markets. Demand side management will affect industrial process chains and the life of every consumer, thereby creating new social challenges. To overcome these challenges the future energy system needs to be much more flexible, which can only be achieved by enhancing the energy grids with smart information technology (IT) solutions to so called "smart grids". Thus, many groups worldwide collected and analyzed data from existing grids (e.g., [9, 11, 15]), built experimental

© Springer International Publishing Switzerland 2015
S. Gottwalt et al. (Eds.): EI 2015, LNCS 9424, pp. 83–94, 2015.
DOI: 10.1007/978-3-319-25876-8_8

platforms including microgrids [7, 10], smart homes [14] or living labs, proposed analysis and visualization strategies [23, 24], established simulation systems [13, 18], investigated different simulation scenarios, and developed concepts and prototypical implementations for smart grids. But most of these research projects concentrate on solutions with only a few components.

At the Karlsruhe Institute of Technology (KIT), an experimental test facility for smart grid related research called Energy Lab 2.0 [2, 8] is on the way to realization. It combines larger-scale components for generation, transformation, and storage of various energy carriers, energy consuming devices and facilities, and an advanced IT communication and computing infrastructure to an integrated smart grid research environment. It is designed to study the future topology of energy grids and energy communication networks, scenarios for the operation of future energy systems, the reliability of future system software, security, safety and privacy issues, the structure and design of control and planning tools, the design of models and simulation software, big data issues, and innovative visualization and interaction methods.

One main part of the IT research infrastructure of the Energy Lab 2.0 (overview in Sect. 2) is the Control, Monitoring and Visualization Center (CMVC) that is described in Sect. 3 of the present article in more detail.

2   Energy Lab 2.0 and SEnSSiCC

The Energy Lab 2.0 is funded by the German Helmholtz Association, the German Federal Ministry of Education and Research (BMBF) and the Ministry of Science, Research and Art (MWK) of the State of Baden-Württemberg in Germany. It will be a large experimental test and simulation field for multi-scale and multi-mode energy system facilities and for testing their operation in various smart grid configurations. On the one hand it partially connects already existing energy related electrical, thermal, and chemical plants and assets to establish integrated technical processes for energy storage and energy conversion as is shown in Fig. 1. On the other end it provides an IT communication, monitoring and simulation infrastructure that further integrates the technical plants into an integrated digital research platform for smart grid research. Other existing technical plants, facilities and consumers in buildings on the KIT campus as well as external facilities and a grid laboratory will be integrated into the digital smart grid research platform as well.

The Smart Energy System Simulation and Control Center (SEnSSiCC) is the core component of the IT infrastructure of Energy Lab 2.0. It contains three main parts:

1. The **Smart Energy System Control Laboratory** (Smart Grid Lab), a separate microgrid test field that can provide physically real equipment and serves as an important link between reality and theory. This *Smart Energy Lab* is a Power Hardware in the Loop (PHIL) field in the 200 kW range and will be built-up in close spatial vicinity to a 1 MW PHIL. It will supply data that help in creating and validating models. Since the test field is galvanically isolated from the public power grid, it allows studying operation points and control strategies under extreme conditions that would not be possible in the public power grid. The design is very flexible; its structure and topology can be varied by digitally controlled switches.

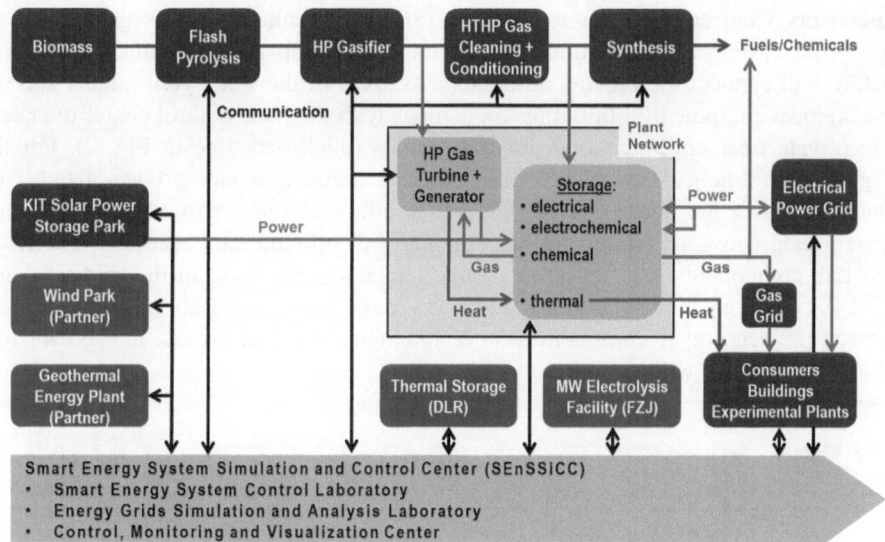

Fig. 1. Energy Lab 2.0 components; *dark gray*: existing facilities; *light gray*: new components (SEnSSiCC, energy conversion and storage plant network); *gray*: external facilities connected in terms of IT to Energy Lab 2.0; *black lines*: communication (Abbreviations: HP – High Pressure; HT – High Temperature; DLR – Deutsche Luft- und Raumfahrt; FZJ – Forschungszen-trum Jülich)

2. The tasks of the ***Energy Grids Simulation and Analysis Laboratory*** comprise the development of software power grid models, their simulation and optimization in combination with the simulation platform that is established by the PHIL compo-nents and various simulation tools. The simulations will be conducted on a microgrid scale as well as on a regional and national scale.
3. The ***Control, Monitoring and Visualization Center*** (CMVC) bundles all commu-nication channels and real facilities, simulation environments, and data repositories into an integrated research environment for planning, control, monitoring, analyzing and visualization of smart grids and their components, and furthermore for evalu-ating future concepts for smart grid utility operation. Section 3 describes this SEnSSiCC component in greater detail.

In addition, cross-sectional activities have been launched for big data, advanced control methods, and reliable, safe and secure software structures [8].

3   Basic Concepts of the Control, Monitoring and Visualization Center

A core component of the Energy Lab is the Control, Monitoring and Visualization Center (CMVC, see Fig. 2). It will be equipped with state-of-the-art commercially available utility control center software and a corresponding communication infrastructure for

Supervisory Control and Data Acquisition (SCADA technology, dark gray boxes in Fig. 2). According to the requirements of the EnergyLab the communication infrastructure will connect the process automation systems of the EnergyLab plants and the measurement equipment of buildings on campus with the local control center over separate private fiber optic networks for data access (thick dark line in Fig. 2). For the integration of Energy Lab plants from external partners secure private networking connections over the Internet (e.g. VPN and similar technologies) will be used. Other separate data networks within the lab building will provide data access to the Smart Grid Lab components (e.g. the process automation systems used in the grid lab), and SCADA communication support for simulators in the Simulation Lab. In supplementary research projects, other communication technologies for data access, like power line communication [5], will also be explored.

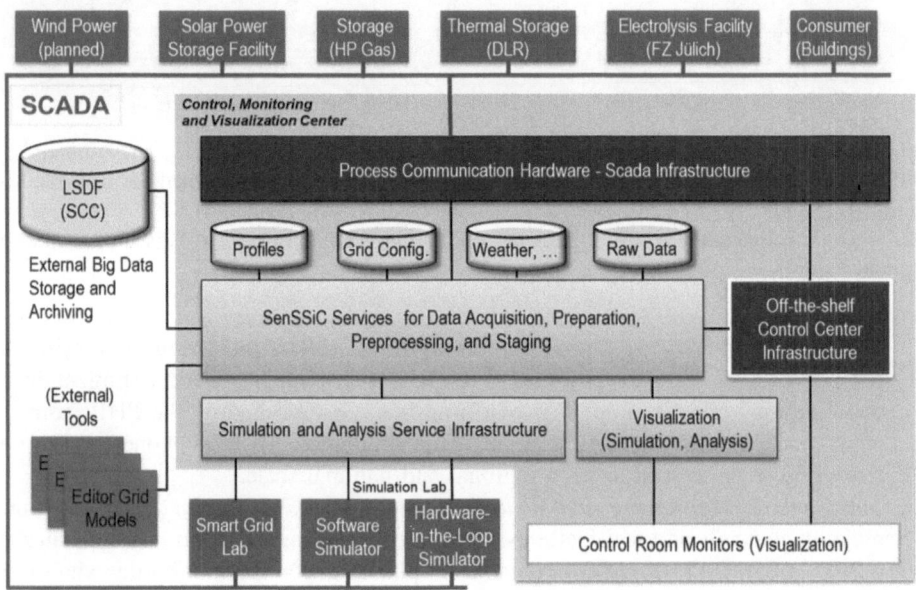

Fig. 2. IT Architecture of the Control, Monitoring and Visualization Center; *gray*: external hardware and software components of Energy Lab 2.0; *light gray*: data repositories (external: LSDF – Large Scale Data Facility) and software modules for data management, simulation management, and visualization; *dark gray*: commercial modules

It is planned, that a commercial software and hardware communication infrastructure facilitates modern Smart Grid communication standards for substations and process automation systems (e.g. technical plant or equipment) to control station communication (i.e. instrumenting IEC 61850 protocols like MMS and Goose, or related, like IEC 60870 TASE.2 or IEC 61970, see [20]). The communication infrastructure should also support older standards via gateways if needed. One of the main research questions addressed by the CMVC is which software architecture and software components are needed for future control centers that have to deal with real "Big Data", i.e. large amounts of data that cannot be stored and analyzed using only a small number of high performance computers. Therefore, besides the commercial control center software that only scales to a limited number of connected substations, a modular and highly scalable software infrastructure for smart grid related research will be designed and implemented as part of the CMVC on top of Big Data technologies (light gray boxes in Fig. 3). This research software infrastructure will primarily run on a virtual computing cluster with several hundred multi-core CPUs and high performance storage arrays. The cluster will be empowered by sufficient data storage and computing capacity so that research oriented software solutions for large scale parallel data analysis (big data analysis) and simulations can be executed and tested on the CMVC hardware infrastructure before they are moved to even bigger Big Data computing environments, such as the computing grid of KIT's Steinbuch Centre for Computing (SCC), one of the largest research-oriented computing and data infrastructures in Europe.

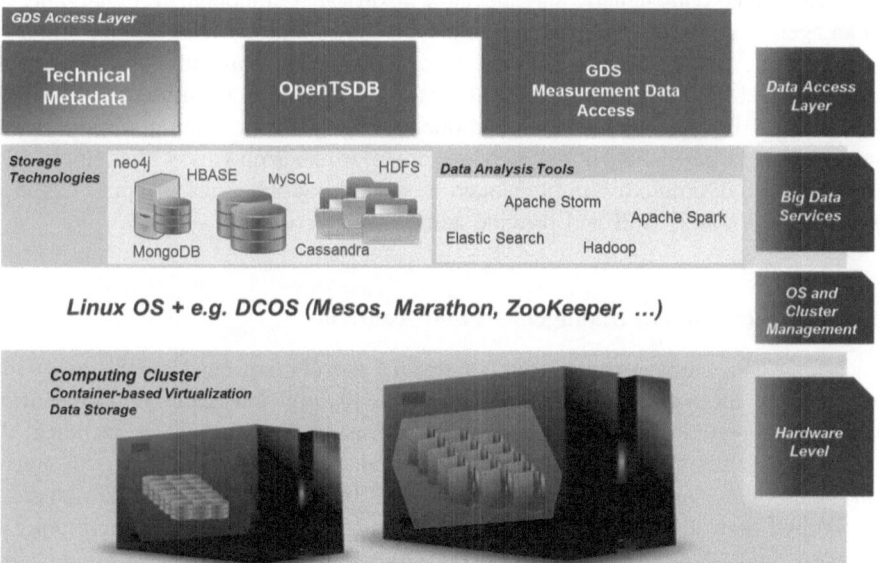

Fig. 3. Components of the Distributed Computing Platform of the CMVC. The Generic Data Services (GDS) in the top service layer will be developed by KIT data management groups [11].

While some bigger utility control centers, especially in the US, have already added such distributed datacenter environments to their control center equipment for additional tasks, e.g. predictive analysis or forecasting, it is the aim of the KIT to explore the usage of such big data software and hardware infrastructures for utility operations from the ground up. Therefore, the research software infrastructure for utility operations of the CMVS will be based on a highly distributed and scalable microservice-oriented architecture. Every microservice and software component is automatically managed and controlled via an underlying distributed computing operating system environment. The cluster will facilitate container virtualization [1], a distributed systems kernel for job and service scheduling, and control technologies (e.g. Docker (www.docker.com), Kubernetes (www.kubernetes.io), Mesos (http://mesos.apache.org/), etc.) as shown in Fig. 3. The distributed computing environment will already contain as elementary software services well known Hadoop components (https://hadoop.apache.org/) for Big Data storage, like the Hadoop File System (HDFS), Hbase or MapR-DB and Cassandra (http://cassandra.apache.org/) as column store databases, and document-oriented or graph-oriented NoSQL databases supported by the environment. Big Data analysis tools like Hadoop MapReduce, Apache Storm, Apache Spark or Drill will also be available.

Operating environments for data intense applications as described above are partially commercially available or can be manually composed out of many existing open source projects. The CMVS will augment and enhance such a computing environment with dedicated implementations of essential microservices needed for future utility operation on top of such infrastructures in order to replace older approaches in current control center software. These services will implement access to different types of smart grid related data, generic and specialized software components for analyzing data and an agent-based simulation infrastructure for the co-simulation of hybrid smart grid configurations. These services are described in more detail in Subsects. 3.2 and 3.3. A modular web-based application infrastructure as described in Subsect. 3.3 provides user interfaces for data visualization and operator access from different kind of devices ranging from large visualizations screens down to smartphones. Other applications (like commercial simulators implemented as desktop applications or apps) can easily access the services as well and provide additional functionality.

3.1   Data Acquisition, Management and Analysis

Concerning data management, there are several aspects that need to be further illustrated. Firstly, there are data from measurements (mainly time series), and event data which are delivered via the communication infrastructure, as well as technical data describing grid configurations, technical plants and equipment, such as measurement units. Common control center software also manages other semi-structured or unstructured data, like log data from equipment, or utility operation handbooks and images of circuit diagrams and layout plans. Other data originate from external sources, like e.g. weather data, or are derived by analysis, manually created by model design, or calculated in simulations and other activities. All these kinds of data have to be stored in the CMVC in a modular fashion (i.e. different services for different kinds of data and data access), and the data services will use different types of database technology as storage engine (e.g. relational databases or different kinds of NoSQL databases). This modularization of the persistence layer of the CMVC research software infrastructure is quite common for larger microservice-oriented applications and termed *polyglot persistence*. It can considerably enhance the performance of larger scale information systems, such as energy data management systems and control center software [17].

In context of the Energy Lab 2.0 and several other projects, technical plants of the Energy Lab and from additional partners of the KIT, buildings, and energy related infrastructure of the KIT campus will be gradually equipped with new Intelligent Electronic Devices (IEDs), new process automation systems or extensions to already existing process automation systems and corresponding communication hardware to provide monitoring data for the CMVC. As already described, it is planned to primarily use equipment which implements newer smart grid communication standards, like the IEC 61850 [12] and associated standards. Where this is not feasible, gateway solutions will be established. For data gathering, software services instrumenting the client part of the IEC 61850 protocol for "control center – substation" communication (MMS), the Inter-Control Center Communications Protocol (ICCP or IEC 60870-6/TASE.2), and other common smart grid data acquisition protocols (i.e. for smart meter data access) by using commercial or open source implementations of these protocols will be developed. These services will then collect measurement and event data and – if available – also technical data from the technical plants, buildings and equipment that are used by the CMVC for monitoring.

All data will be forwarded, stored and accessed through distinct data services that will be tuned for high performance and parallel and asynchronous operation. Most of these data services will likely use NoSQL database technology already available on the distributed computing cluster as storage engine. In first tests, an installation of openTSDB (open Time Series Database) on top of the Hadoop database HBase was evaluated for real-time storage of measurement data. As described in [16] openTSDB can handle large amounts of time series data sets easily and provides good performance both for real-time storage and data access. As storage engine for technical metadata document-oriented databases, such as MongoDB, were evaluated. They provide a flexible schema-less data model to store technical metadata as JSON objects which can easily combine structured, semi-structured and unstructured metadata fields. Because of the schema-less design of such databases, new types of metadata objects can be easily added or augmented by further information.

Another document-oriented database technology, ElasticSearch, is already used in several projects at the KIT for providing advanced search functionality for large scale information systems [19]. ElasticSearch will likely be integrated into the CMVC data infrastructure as main infrastructure for search driven access to technical metadata objects and binary assets. In another test, the possibilities of ElasticSearch for serving as a central access point to event and log based data was explored. The test shows, that ElasticSearch together with other databases and analytics tools, like Apache Spark or Apache Storm, can provide a good platform for near-real-time access to semi-structured event and log data (see also [4]).

The data services implemented for the CMVC will provide easy to use REST-based (Representational State Transfer) APIs for data access that can be used out-of-the-box for data analysis, data transformation, and simulation services. For archiving, the CMVC data services will be connected to the external Large Scale Data Facility (LSDF at SCC [6]), where mainly simulation and data analysis results and large datasets like raw data time series can be stored for longer time intervals.

The data services will provide dedicated input and output source interfaces for common data analysis frameworks (e.g., Apache Hadoop/MapReduce, Apache Spark, Apache Storm) [22] installed on the distributed computing operating system of the

CMVC that allow direct access to the data services via REST-based data (streaming) APIs (e.g. "Spouts" in the case of Apache Storm or resilient distributed datasets in the case of Apache Spark). These analysis frameworks can then be used by researchers to implement and evaluate different kinds of data-driven analysis algorithms, e.g. for data-driven predictive analysis, forecasting or real-time event analysis. Also standalone analysis tools, like R or Matlab can be connected to the data infrastructure as well. The whole infrastructure will adhere to a distributed form of streaming concept, like the well-known concept of pipes in UNIX-based systems, where data services can be accessed as data streams that are fed into a distributed analysis chain that sends these data streams to other data services for storage. The whole data and analysis infrastructure should allow scientists to easily explore data, prototypically implement analysis algorithms using one of the provided frameworks, and access and explore results with additional data visualization tools (see Subsect. 3.3).

3.2   Simulation Service Infrastructure

The simulation service infrastructure will enable scientists to perform co-simulations of arbitrary smart grid configurations. It will facilitate scientists to combine live data from Energy Lab plants or equipment of the grid laboratory with data from various software-only or power hardware-in-the-loop simulators (Smart Grid Lab and PHIL). Each simulator, plant or equipment will be seen by others and/or other control center software through a communication interface that connects to the smart grid communication and SCADA infrastructure.

For grid components that are only simulated in software a generic frontend communication server preceding each simulator instance will implement the IEC 61850 protocols. At the backend it can be configured to implement a simulator-specific device interface that allows exchanging measurement and control data between the simulator and the communication and SCADA infrastructure of SEnSSiCC. The frontend communication server will also implement a communication interface for simulating a connection to the simulated electricity grid (i.e. a connection to AC or DC lines). If the simulated grid component needs to be connected to the electricity grid, this interface can be configured to exchange AC and DC information on the virtual electricity grid connection. A third communication interface implemented by the frontend communication server allows for exchanging metadata about external influences, such as weather or market conditions, or feeds data or simulation profiles into the simulator. And finally, a control channel implemented in each frontend communication server will allow starting, stopping or monitoring simulator instances.

For interfacing to a hardware-in-the-loop simulator the frontend communication server typically does not need to implement the electricity grid connection because it is already in-the-loop. For a real plant or other equipment instrumented by IEDs, the connection to the communication infrastructure is already provided by the IED itself. During a simulation run, each software simulator will run on the computing cluster in a virtualized container or on a virtual machine (if needed) that provides the necessary operation system environment (Linux or Windows).

3.3   Data Visualization and User Interaction

Client applications for data visualization and operator interaction are a vital part of the CMVC. They allow operators to monitor, plan and control the grid. Until now, control center software often uses a client-server architecture approach where the operator interface is composed of several (Windows) desktop applications. Probably, this is also the case for the commercial control center software integrated into the CMVC. In microservice-oriented architectures as realized for the research oriented software infrastructure of the CMVC, service interfaces can be easily used by different types of client applications, e.g. desktop applications, mobile apps and web-based applications. Because the research oriented part of the control center is used for the prototypical implementation and evaluation of new smart grid concepts, a natural approach for many types of such experimental user interfaces will be modular web-based applications with responsive and adaptable user interface components. This will allow componentizing the user interface and deploying it on different kind of hardware ranging from large display walls down to tablets or even smartphones. For the mere visualization of data and results of analysis frameworks as mentioned in Subsect. 3.1, there are already many out-of-the-box usable web-based analysis and visualization frameworks (see Fig. 4), that can be easily integrated into the tool stack of the control center. These allow exploring data quickly without programming and can even visualize data provided as the result of preceding analysis runs. Visualizations of such tools can be easily embedded into other web applications and combined with other information components.

By adding a web-based portal server to the control center portfolio, it is very easy to aggregate and combine different visualizations and information display components into modular composed web pages (e.g., dashboards like information pages as shown

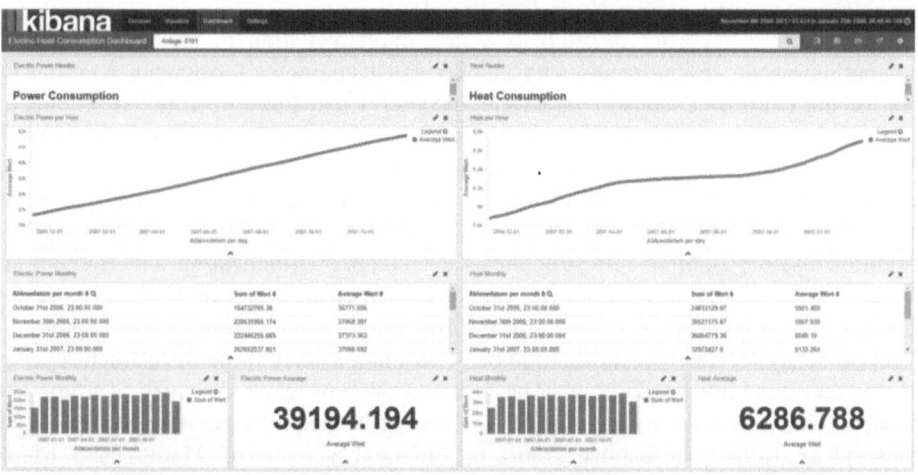

Fig. 4. Statistics about electricity power and heat consumption of a building at Campus North of the KIT. Energy-related management data of Campus North buildings are already stored in an ElasticSearch infrastructure and can be analyzed and visualized without programming with Kibana (https://www.elastic.co/).

in Fig. 4) through the use of portal technology. The user interfaces for utility operation can then completely be built up by composing such modular web pages.

Web technology and mobile applications also provide a good environment for testing and exploring new types of interactions for control center operation. The options range from the use of touch-screen based interactive tables, over using smartphones or tablets as remote controls for bigger display walls for performing control operations, to contact-free interaction with larger display and interaction hardware via motion detection or even voice based interaction. But not all new interaction mechanisms will be adequate for safe utility operation. The interaction and display hardware and software acquired for the CMVC will allow implementing and exploring different kinds of new interaction mechanisms to find out which of these new interaction possibilities will be useful for future control center operation.

4 Status and Outlook

The KIT has already performed some essential work on the data platform of the control center and the use of Big Data technology. For measuring current and power of electricity lines with a high sample rate of 12800 samples per second, an Electrical Data Recorder (EDR and generic data services for storing and retrieving these measurement data have been developed) [11]. It sends its data in a special XML sample file format in larger chunks to a data service that archives the samples in the Hadoop file system infrastructure of the SCC. The same data recorder aggregates measurement data every second and sends the aggregated power data as status events to another data service which stores this data in an ElasticSearch infrastructure. As already mentioned, ElasticSearch is also used in first tests for storing and accessing energy monitoring data of all buildings of the Campus North (see Fig. 4).

A smaller test cluster is setup as a first implementation of the Control Center computing cluster. Experiments with Docker and Kubernetes for container virtualization of the control center software were successful and this infrastructure is already used as described above. Furthermore, Hadoop runtime infrastructures and NoSQL databases were explored as already discussed in Subsect. 3.3. A key result of these tests is that the mentioned tools work well in the control center application context, however the manual setup of Hadoop, Docker und Kubernetes runtime infrastructure is tedious. Additionally, the runtime infrastructures for container virtualization and Hadoop are not fully aligned (e.g. a Hadoop job cannot be started in a Docker container per se). Currently, a lot of work is ongoing in the Big Data tools community to integrate the Hadoop ecosystem fully with distributed scheduling and runtime infrastructures, e.g. Mesos and container technology. These efforts already resulted in new startups and products, like the early access version of DCOS (Datacenter Operating System) from Mesosphere (https://mesosphere.com/) or the next version of MapR from MapR Technologies (https://www.mapr.com/), a commercial Hadoop distribution which currently is extended to provide also full support for Mesos and container virtualization. These environments will likely provide easy usage and administrate environments for running Big Data infrastructures and microservice-based applications. The KIT research group responsible for the CMVC has already access to these technologies for early testing which will be done over the next few months.

For the implementation of the IEC 61850 data access clients, several open source and commercial libraries implementing the protocols are available which are currently evaluated. The requirements for integrating technical infrastructures into the communication infrastructure of the control center were clarified and the technical equipment will be gradually integrated into the communication infrastructure of the control center as the hardware becomes available. Finally, there are ongoing efforts involved in integrating the analysis frameworks into the platform and in detailing the simulation services.

References

1. Bernstein, D.: Containers and cloud: from LXC to docker to kubernetes. IEEE Cloud Comput. **1**(3), 81–84 (2014)
2. Energy Lab 2.0. (Teil-) Antrag an das Ministerium für Wissenschaft, Forschung und Kunst Baden-Württemberg (2015)
3. Fang, X., Misra, S., Xue, G., Yang, D.: Smart grid – the new and improved power grid: a survey. IEEE Commun. Surv. Tutor. **14**, 944–980 (2012)
4. Fischer, F., Keim, D.A.: NStreamAware: real-time visual analytics for data streams to enhance situational awareness. In: Proceedings of the Eleventh Workshop on Visualization for Cyber Security, pp. 65–72. ACM (2014)
5. Galli, S., Scaglione, A., Wang, Z.: For the grid and through the grid: the role of power line communications in the smart grid. Proc. IEEE **99**(6), 998–1027 (2011)
6. Garcia, A.O., Bourov, S., Hammad, A., et al.: The large scale data facility: data intensive computing for scientific experiments. In: Proceedings of the IEEE International Symposium on Parallel and Distributed Processing Workshops and Ph.D. Forum, pp. 1467–1474 (2011)
7. Ghareeb, A.T., Mohamed, A.A., Mohammed, O.A.: DC microgrids and distribution systems: an overview. In: IEEE Power & Energy Society General Meeting, pp. 1–5 (2013)
8. Hagenmeyer, V., Cakmak, H.K., Düpmeier, C., Faulwasser, T., Isele, J., Keller, H.B., Kohlhepp, P., Kühnapfel, U., Stucky, U., Mikut, R.: Information and Communication Technology in EnergyLab 2.0. In: Proceedings of the Energy Science Technology, Karlsruhe (2015)
9. Hammerstrom, D.J., Ambrosio, R., Brous, J., et al.: Pacific Northwest GridWise Testbed Demonstration Projects, Part I. Olympic Peninsula Project. Pacific Northwest National Laboratory (PNNL), Richland (2007)
10. Kotsampopoulos, P., Kapetanaki, A., Messinis, G., Kleftakis, V., Hatziargyriou, N.: A power-hardware-in-the-loop facility for microgrids. Int. J. Distrib. Energy Resour. Technol. Sci. Publishers **9**(1), 89–104 (2013)
11. Maaß, H., Cakmak, H.K., Bach, F., Mikut, R., Harrabi, A., Süß, W., Jakob, W., Stucky, K.-U., Kühnapfel, U.G., Hagenmeyer, V.: Data processing of high rate low voltage distribution grid recordings for smart grid monitoring and analysis. EURASIP J. Adv. Sig. Process. **2015**, 21 (2015)
12. Mackiewicz, R.E.: Overview of IEC 61850 and benefits. In: IEEE PES Power Systems Conference and Exposition (PSCE 2006). IEEE (2006)

13. Müller, S.C., Georg, H., Rehtanz, C., Wietfeld, C.: Hybrid simulation of power systems and ICT for real-time applications. In: Proceedings of the 3rd IEEE PES International Conference and Exhibition on Innovative Smart Grid Technologies (ISGT Europe), pp. 1–7 (2012)
14. Paetz, A.-G., Becker, B., Fichtner, W., Schmeck, H.: Shifting electricity demand with smart home technologies–an experimental study on user acceptance. In: Proceedings of the 30th USAEE/IAEE North American Conference, vol. 19, p. 20 (2011)
15. Pignati, M., Popovic, M., Barreto Andrade, S., et al: Real-time state estimation of the EPFL-campus medium-voltage grid by using PMUs. In: Proceedings of the Sixth Conference on Innovative Smart Grid Technologies (2015)
16. Prasad, S., Avinash, S.B.: Smart meter data analytics using OpenTSDB and hadoop. In: 2013 IEEE Innovative Smart Grid Technologies - Asia (ISGT Asia), pp. 1–6, 10–13 November 2013
17. Prasad, S., Avinash, S.B.: Application of polyglot persistence to enhance performance of the energy data management systems. In: 2014 International Conference on Advances in Electronics, Computers and Communications (ICAECC), pp. 1–6. IEEE (2014)
18. Rohjans, S., Lehnhoff, S., Schutte, S., Scherfke, S., Hussain, S.: mosaik - A modular platform for the evaluation of agent-based smart grid control. In: Proceedings of the Innovative Smart Grid Technologies Europe (ISGT EUROPE), pp. 1–5 (2013)
19. Schlachter, T., Düpmeier, C., Kusche, O., Schmitt, C., Schillinger, W.: Towards a search driven system architecture for environmental information portals. In: Denzer, R., Argent, R. M., Schimak, G., Hřebíček, J. (eds.) ISESS 2015. IFIP AICT, vol. 448, pp. 351–360. Springer, Heidelberg (2015)
20. Sato, T., et al.: Smart Grid Standards: Specifications. Requirements and Technologies. Wiley, Singapore (2015). ISBN 978-118-65369-2
21. Scholz, R., Beckmann, M., Pieper, C., Muster, M., Weber, R.: Considerations on providing the energy needs using exclusively renewable sources: energiewende in Germany. Renew. Sustain. Energy Rev. **35**, 109–125 (2014). Elsevier
22. Shukla, R.K., Pandey, P., Kumar, V.: Big data frameworks: at a glance. Int. J. Innovations Adv. Comput. Sci. IJACS **4**(1), 169–175 (2015)
23. Waczowicz, S., Klaiber, S., Bretschneider, P., Konotop, I., Westermann, D., Reischl, M., Mikut, R.: Data mining to analyse the effects of price signals on household electricity customers. at -Automatisierungstechnik **62**, 740–752 (2014)
24. Zhang, P., Li, F., Bhatt, N.: Next-generation monitoring, analysis, and control for the future smart control centre. IEEE Trans. Smart Grid **1**, 186–192 (2010)

Building Energy Management in the FZI House of Living Labs

Birger Becker$^{(\boxtimes)}$, Fabian Kern, Manuel Lösch, Ingo Mauser$^{(\boxtimes)}$,
and Hartmut Schmeck

FZI Research Center for Information Technology,
Haid-und-Neu-Str. 10-14, 76131 Karlsruhe, Germany
{bbecker,kern,loesch,mauser,schmeck}@fzi.de
http://www.fzi.de, http://www.house-of-living-labs.de

Abstract. The *FZI House of Living Labs* is a research and demonstration environment that facilitates interdisciplinary research, development, and evaluation in real-life scenarios. It consists of various *Living Labs* addressing different research topics. In the *Living Lab smartEnergy*, solutions for the energy system of the future are investigated. For this reason, the whole FZI House of Living Labs has been equipped with building automation, distributed generation, thermal and electrical storage, and technologies that enable the flexibilization of energy supply and demand. The equipment, among others, includes a photovoltaic and battery storage system, a micro combined heat and power plant, and an adsorption chiller. A building energy management system was developed that integrates various communication technologies, and hence enables monitoring, data recording, visualization, and the integrated optimization of the devices and systems. This way, flexibilities can be utilized with regard to different optimization goals such as an increased self-consumption, or the provisioning of grid-supporting services.

Keywords: Energy management · Smart building · Smart home · Building automation · Demand side management

1 Introduction and Motivation

Governments worldwide set the goal to reduce greenhouse gas emissions. The resulting energy transition with an increasing share of renewable energy resources comes along with an intermittent and highly fluctuating energy supply. Hence, as demand and supply within the grid always have to be balanced, one of the key challenges is the efficient utilization of load flexibility. Due to the high energy demand in buildings, *Building Energy Management Systems* (BEMS), which are sometimes also called *Building Operating Systems* [8], bear a great potential for adapting the building's energy load to the global grid state, as well as for providing appropriate ancillary services in future smart grids.

Since 2011, such a BEMS is developed and deployed in the research and demonstration environment *FZI House of Living Labs* (HoLL), which facilitates interdisciplinary research, development, and evaluation in various real-life

© Springer International Publishing Switzerland 2015
S. Gottwalt et al. (Eds.): EI 2015, LNCS 9424, pp. 95–112, 2015.
DOI: 10.1007/978-3-319-25876-8_9

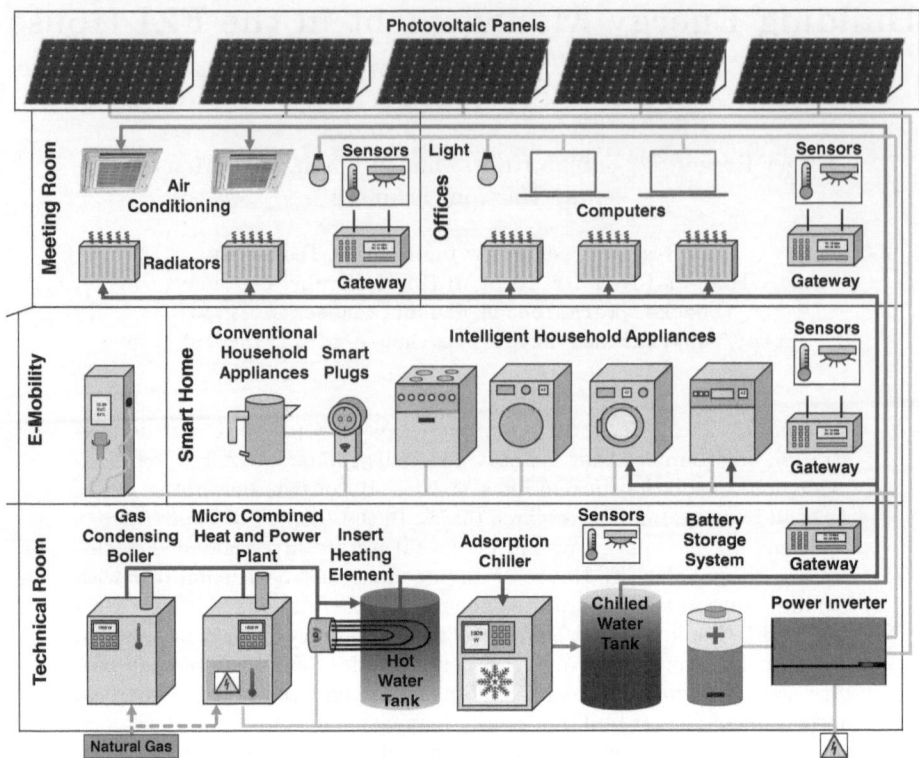

Fig. 1. Structural plan of electrical and thermal equipment of the HoLL

scenarios [17] at the *FZI Research Center for Information Technology* in Karlsruhe, Germany. At the same time, the HoLL also serves as office building for the FZI. In the *Living Lab smartEnergy*, our research focuses on efficient information and communication technologies (ICT) for the integration of heterogeneous components into the BEMS, as well as on optimization strategies and algorithms for energy systems and user interaction interfaces.

With the objective of providing a durable and innovative research lab for energy applications, the HoLL is equipped with state-of-the-art technology that is extended continuously. As the overarching research objective is to provide flexibility for the energy grid, many of the devices and systems provide communication interfaces. The standardized integration of different communication protocols is a major focus of the research in the HoLL. Therefore, it is equipped with heterogeneous communication systems. Since the integrated optimization of various energy carriers is another research focus, energy flows for electric installations, heating, cooling, and natural gas are measured, analyzed, and controlled.

In this paper, the parts of the research environment HoLL that focus on private households and commercial buildings as well as investigated research topics are presented. An overview of the energy equipment and energy flows within the HoLL is illustrated in Fig. 1. The dashed orange lines indicate the supply with natural gas, yellow lines corresponds to electrical energy flows, red and blue lines represent thermal energy flows of hot and chilled water. A large number of different appliances, devices, and building automation technologies provides a suitable research platform for various aspects such as standardized integration, interoperability, smart grid capabilities, and energy management methods. Based on this platform, optimization strategies are developed to cope with different energy management objectives, such as grid support, economic profit, or self-consumption. Due to significant interdependencies of electrical and thermal energy supply, consumption, and storage, which are more closely described in the following sections, their joint management and optimization is done in one BEMS. As the BEMS is integrated into the real operation of the building, valuable data and experience with real-life application are gathered.

Projects that utilize the HoLL for research on energy topics include publicly funded research projects, cooperations with companies of the automation, automotive, electrical equipment, energy, household appliance, and software industry, the *Energy Lab 2.0* of the *Karlsruhe Institute of Technology* (KIT), and the project *Storage and Cross-linked Infrastructures* of the *Helmholtz Association*.

This paper is structured as follows: Sects. 2 and 3 provide an overview of the electrical and thermal equipment that is installed in the HoLL; the installed building automation systems are described in Sect. 4. The software architecture, selected optimization strategies of the BEMS, and concepts for user interaction are depicted in Sect. 5. Section 6 presents exemplary research that is based on the HoLL as well as lessons learned when setting up the environment. An overview of similar research environments is given in Sect. 7. Finally, Sect. 8 summarizes the main aspects and gives an outlook on future work.

2   Electrical Equipment

Various electrical suppliers, consumers, storage systems, and electric vehicles (see Fig. 1 and Table 1) are integrated into the real building operation. Due to the research aspects mentioned, a wide range of devices and systems of different manufacturers have been selected. Each of these devices and systems provides a communication interface, resulting in a heterogeneous system infrastructure.

The *photovoltaic* (PV) system provides up to 15.1 kW using 108 *photovoltaic panels* which are installed on the rooftop. A *battery storage system* with a capacity of 30 kWh is installed. Three *power inverters* convert direct current (DC) provided by the PV panels and the battery storage system into alternating current (AC). A grid outage can be bridged by temporarily providing an islanded AC grid via the power inverters. To increase the autonomy time, the island grid provides power to selected rooms only. The power inverters have an USB communication interface, which enables read and write access to the system. To provide a more generic *Representational State Transfer* (REST) interface for read (e.g., currently generated power) and write (e.g., dynamic control of power factor) access, we have installed an embedded system. This way, the BEMS is able to control the PV system for increasing the consumption of locally generated energy and for providing ancillary services to the power grid, e.g., improving voltage quality or compensating unbalanced phases. Using data recorded in the HoLL, this aspect has been more closely investigated in [5].

Additionally, a gas-fired *micro combined heat and power* unit (µCHP) providing electrical and thermal energy has been installed. The major advantage of the CHP technology is a high overall efficiency by using 12.5 kW waste heat for heating purposes when generating 5.5 kW electrical power. The integration of thermal storage tanks allows for electricity generation by the µCHP unit even when the thermal energy is not consumed at that time (see Sects. 3 and 5).

An electrical *insert heating element* (IHE) is installed in the hot water tank and coupled to the PV system. The objective is to convert surplus electrical energy generated by the PV system into thermal energy instead of feeding the surplus energy into the grid. For this purpose, the PV system has been equipped with additional measurement technology that communicates directly with the IHE.

A major research aspect concerning BEMS is the development of strategies and algorithms to detect and exploit electrical demand and supply flexibility of buildings. To investigate the potential of electrical load management in private households, a *smart home* has been set up within the *Living Lab smartHome/AmbientAssistedLiving*. It is equipped with numerous household appliances of different manufacturers and used by employees of the FZI. Most of the appliances are equipped with communication modules providing an appropriate external communication interface (*intelligent household appliances*). Alternatively, *smart plugs* are installed at the electricity connection of the appliances. Being put between the wall socket and the device plug, they allow to turn connected devices on and off, and to measure their energy consumption. This way, devices that are able to continue their programs at the point where they have been interrupted (fail-over capability) can also be controlled by the BEMS, although not being equipped with communication capabilities by the manufacturer. Besides measuring various devices, this is in particular used to retrofit

Table 1. Technical specifications of selected electrical devices

Device	Specification	Manufacturer and type
PV panels	Electrical power: 15.1 kW$_{peak}$	108× *Würth Solar WSF0002E140*, CIS
Battery storage	Electrical capacity: 30 kWh	3× *BAE AK40012*, VRLA
Inverters for PV and battery	Max. electrical power: 5 kW per phase	3× *Nedap PowerRouter PR50SB*
µCHP unit	Thermal power: 12.5 kW, Electrical power: 5.5 kW	*Senertec Dachs G 5.5 standard*
Electric vehicles	Max. charging power: 22 kW, Capacity: 15.1 kWh	*Smart Fortwo Electric Drive*
	Max. charging power: 3.6 kW, Capacity: 40.0 kWh	*Peugeot 3008* (modified)

conventional household appliances. The communication modules and the smart plugs are communicating with a dedicated gateway (see Sect. 4). Their integration into the BEMS enables the system to monitor the usage, the state, and the energy profile for each appliance. The research in the smart home focuses on standardized communication and integration including strategies for the automatic control and optimization of appliances.

To take advantage of the energy feed-in and consumption tariffs, we have to distinguish between power of the PV system and power of the μCHP unit, as well as grid feed-in and self-consumption. Therefore, we installed a complex metering infrastructure consisting of several (partial two-way) meters, which enables the measurement of the different electrical power flows. In a second step, we integrated the meters into the BEMS using a standardized optical meter interface and the *Device Language Message Specification* (DLMS) protocol.

3 Thermal Equipment

Space heating and air conditioning (see Fig. 1 and Table 2) are done by a system that comprises a gas-fired μCHP unit, an adsorption chiller, a gas-fired condensing boiler, the IHE, as well as storage tanks for hot and chilled water. Such a kind of thermal system is called *trigeneration* or *combined cooling, heat, and power* (CCHP). It combines a μCHP unit, i.e., *cogeneration*, with an adsorption chiller that produces chilled water. Additionally, hot water is also produced by the condensing boiler and the IHE. Generation and consumption of chilled respective hot water are decoupled by the storage tanks.

Two ceiling-mounted cassettes use the chilled water to air-condition a meeting room, which is used by employees of the FZI. The adsorption chiller is mainly powered by hot water supplied by the hot water tank. Bookings of the meeting room trigger air-conditioning requests for the preconditioning in advance to

Table 2. Technical specifications of selected thermal devices

Device	Specification	Interface	Manufacturer and type
Adsorption A/C	Cooling power: 9 kW	Digital I/O	*InvenSor LTC 09*
Gas-fired μCHP unit	Thermal power: 12.5 kW, Electrical power: 5.5 kW	Digital I/O	*Senertec Dachs G 5.5 standard*
Gas-fired condensing boiler	Thermal power: 95 kW	0–10 V	*Elco THISION L 100*
Electric insert heating element	Electrical power: 0, 0.5 ... 3.5 kW	Modbus/Serial	*E.G.O. Smart Heater*
Hot water tank	3250 l	–	Custom-made tank
Chilled water tank	3000 l	–	Custom-made tank
System controller	–	Modbus/TCP	*SolarNext chillii*
Heat flow metering system	–	Meter-Bus (M-Bus)	*Aquametro AMBUS Net*

meetings in the room, when the past outdoor temperatures exceed certain temperature thresholds. The bookings are extracted from the room's calender on the *Microsoft Exchange Server*. An *Energy Management Panel* (see Sect. 5.2) in the meeting room enables the visualization of the bookings and the air conditioning states as well as the adjustment of the room set temperature. The BEMS schedules the operation of the μCHP to provide hot water for the adsorption chiller, the operation of the adsorption chiller to provide chilled water, and the actual air conditioning of the room (see Sect. 5.1).

The heating system utilizes numerous sensors and actuators, such as temperature sensors and heating valves. Some of the sensor data is collected by a subsystem (*SolarNext chillii* system controller), which also controls the release signals for the μCHP unit, the condensing boiler and the adsorption chiller. This subsystem works autonomously when no signal is provided by the BEMS. Comprehensive data of the heating system is retrieved by the BEMS using *Modbus/TCP* to be included into the global optimization of the building's energy management. Additionally, the μCHP unit is equipped with a communication interface that provides a *REST service* for collecting data and receiving control signals directly from the BEMS. On the consumption side, building automation systems are integrated into the BEMS to control the heating for each room of the building individually (see Sect. 4).

In order to gain a more detailed view of the energy flows within the heating and cooling system, we installed six additional heat meters for hot and chilled water. Each of the meters consists of an ultrasonic flow measuring unit and two temperature sensors (flow and return). The meters measure current thermal power flows and thus associated energy demands and supplies. They are connected to a central gateway via *Meter-Bus* (M-Bus), which is integrated into the BEMS using its web service interface. In addition to the information about the energy flows, the basic volume flow data is used for predictive maintenance.

4   Building Automation

A heterogeneous system landscape with multiple building automation technologies has been installed to gain insights into many different existing implementations and their integration, as well as to derive concrete requirements for BEMS. Table 3 provides an overview of the automation technologies deployed in the HoLL. Their usage and integration into the BEMS is subsequently described.

4.1   Wireless Technologies

Bluetooth Smart [6] focuses on radio communication with devices characterized by limited computational power and limited energy supply. It is used to communicate with household appliances that are equipped with a manufacturer-provided communication module providing an external interface. This allows interfering with internal control mechanisms by deferring the appliance operation cycle or even interrupting it. The Bluetooth gateway is implemented on a

Raspberry Pi that is enhanced with a USB-connected Bluetooth radio transmission module.

EnOcean [10] is a communication technology for computationally limited, low-power devices. To operate and communicate, EnOcean devices usually leverage environmental energy wherever possible, such as ambient light and temperature differences. EnOcean is used for controlling the room heating with heating valves for radiators and temperature sensors. The gateway to interact with the BEMS is implemented on a Raspberry Pi that is enhanced with a proprietary general-purpose input/output (GPIO) module for EnOcean radio transmission.

ZigBee [39] focuses on small, low-power devices and allows to increase the range of the radio transmission by passing data through a mesh network of intermediate devices. We use ZigBee smart plugs to monitor the power consumption of household appliances, and for enhancing conventional devices into controllable devices (see Sect. 2). The gateway for interaction with the BEMS is based on a proprietary embedded system comprising a ZigBee radio antenna and provides an EEBus interface.

EEBus [9] is a standardization effort working on a common interface for various devices while relying on existing communication technologies. Its goal is to abstract and translate from various existing standards to a common interface. In our environment, EEBus is used for the interaction with the ZigBee smart plugs and appliances.

Additionally, various *proprietary standalone solutions* are deployed to gain deeper insights via evaluations of technical aspects in real-world scenarios. One

Table 3. Selection of technologies used for building automation and device communication in the HoLL

Technology	Communication link	Integration into BEMS (endpoint)	Usage examples
Bluetooth Smart	Radio (2.4 GHz)	*Raspberry PI* with *Bluetooth* USB module	Household appliances, beacons
EnOcean	Radio (868 MHz)	*Raspberry PI* with *EnOcean* GPIO module	Heating valves, lighting, switches, temperature sensors, window sensors
ZigBee	Radio (2.4 GHz)	Proprietary embedded systems	Household appliances, smart plugs
HabiTeq	Wired	*HabiTeq CTD Controller*	CO_2 sensors, heating valves, light sensors, motion sensors, shutters, switches, temperature sensors, window sensors
KNX	Wired	*tebis KNX Domovea Server*	Fire detectors, heating valves, motion sensors, shutters, switches, temperature sensors, window sensors
EEBus	–	*E.G.O. Smart Gateway*	Household appliances, smart plugs
REST over HTTP	–	Direct connection to local Ethernet	Heat meters, μCHP unit, infrared transceivers

example is a *RWE Smart Home* [33] installation which is used to control lighting in one room and heating in two office rooms. Due to the closed eco-system, the lack of an open local interface, and the limited gains that would be possible, it is not integrated into the BEMS. Furthermore, *en:key* [20], an EnOcean-based solution from *Kieback & Peter*, is used for heating control in the *Living Lab smartEnergy* and in another office room. Based on a motion sensor and a learning algorithm, this system uses historical data to predict occupancy and to heat up the room only when required. Other examples for proprietary stand-alone solutions are *Plugwise* [32] smart plugs, which can be accessed using a USB dongle and a proprietary Plugwise control software. Although the communication protocol is proprietary and closed, a third-party *Java* library is available and used to enable communication between the BEMS and the Plugwise smart plugs.

4.2   Wired Technologies

HabiTeq [16] is a proprietary wired automation technology that is offered by *General Electric* and has originally been developed by *Qbus*. It is used in two office rooms for managing wall switches, lighting, temperature and CO_2 sensors, shutters, and heating valves. HabiTeq can be operated as a standalone solution, as a hardware server enforces the interplay of the different components following predefined rules that have to be specified using a proprietary configuration software. At the same time, the hardware server allows also to access the devices using REST, hence allowing for an integration of the proprietary automation solution into the BEMS.

KNX [21] is a widespread building automation bus. We use a twisted-pair KNX installation in the smart home, which comprises different KNX components, such as controllable shutters, temperature and window sensors. In general, static rules for the interplay between different components can be configured using a client software. In our setup, the additional deployment of a proprietary KNX product, *tebis KNX domovea* from *Hager* that provides a touch panel for direct end-user interaction, and a hardware server make the KNX bus accessible over IP. This way, the hardware server is connected to the BEMS, hence allowing to access, e.g., temperature information and window states.

Additionally, single devices offer an HTTP interface with REST. In these cases, no additional gateways are required for the integration into our BEMS. One example is the μCHP unit, which provides a proprietary REST interface allowing to read various values regarding the μCHP usage and to explicitly trigger a start or stop. Furthermore, several heat meters are installed to study the energetic building performance and to enable thermal predictions. These heat meters are also equipped with a REST interface, which is directly integrated into the BEMS.

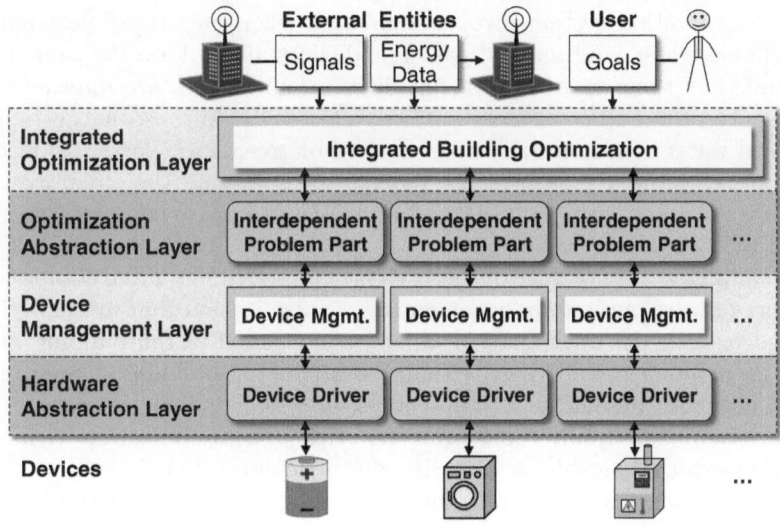

Fig. 2. Overview of organic smart home and its architecture

5   Energy Management and Interaction

This section focuses on the energy management, which is based on the introduced technical equipment as well as the deployed communication technologies and pre-existing automation solutions. It is realized using a building energy management system—the *Organic Smart Home*—and a human machine interface—the *Energy Management Panel*.

5.1   Building Energy Management System

In order to be able to optimize energy flows, an appropriate energy management system that allows for the integration and optimization of the multitude of devices has been developed at FZI and KIT: the *Organic Smart Home*[1] (OSH) [2,27]. It has first been deployed to the *Energy Smart Home Lab* at the KIT [3], where it has been evaluated in multiple evaluation phases [30]. Major advantage of the OSH is its applicability to both, real world energy management and simulations of buildings with different sets of devices in diverse scenarios, enabling the development and testing of control and optimization functionality in simulations before applying them to productive systems. The OSH has already been used to study diverse scenarios of optimizing, e. g., the operation of appliances [2,25], heat-pumps [24], trigeneration [26], and electric vehicles [29]. In order to reach a detailed co-simulation of electrical and thermal energy flows it also has been coupled with the thermal simulation framework TRNSYS [23].

---

[1] http://www.organicsmarthome.com.

The trigeneration system involves four main energy carriers: electricity, hot water, chilled water, and natural gas. The devices all work on the same storage tanks and thus their consumption, production, and states are interdependent. Consequently, the OSH considers not only electricity, but also natural gas, hot and chilled water consumption, and emissions of greenhouse gases. Additionally, the inverters of the PV system are capable of adjusting their reactive power to provide ancillary services [5], based on signals that are communicated from external entities to the OSH.

The simplified architecture of the OSH is depicted in Fig. 2 and comprises several layers for abstraction, management, and optimization that are more closely described in [27]. For optimization, a simplified model of the building and the devices is included in the OSH. Usually, systems for building simulation that focus on thermal simulation use time steps on a scale of several minutes [7]. In contrast, when additionally considering electricity, the optimization with respect to variable external signals (e.g., tariffs and load limitations), user preferences (e.g., specified degrees of freedom), and goals (e.g., cost minimization), requires the utilization of shorter time steps in order to take account of short-time consumption and production peaks as well as the actual self-consumption of the generated power [38]. Therefore, the OSH is able to use time steps with a resolution of one second, thus allowing to investigate the optimum time granularity for optimizations [26].

Optimization of a time horizon of several hours would result in mixed integer linear—or nonlinear when also respecting non-linearities—programming with thousands of constraints and variables, which is usually not solved within adequate time on computers with limited resources [1,34]. Additionally, the execution time of the optimization algorithm is crucial, because frequent rescheduling is quite likely and a quick response to user interaction is desirable [30]. Additionally, solving the optimization problem to optimality is only possible when having complete information about future energy flows, which cannot be obtained ex ante. Thus, generating approximate solutions by a heuristic that allows for frequent rescheduling in varying setups promises to be of better use for productive energy management. Therefore, we decided to use a meta-heuristic—an *Evolutionary Algorithm*—with dynamic formulation of the problem instances at runtime of the system to optimize the system heuristically [2,26].

5.2   Energy Management Panel

For covering the entire potential of BEMS in real-life scenarios, the user acceptance of the system plays an important role. On the one hand, most users want a system running automatically without requiring much user interaction. On the other hand, users often want to understand the actions that have been executed automatically by the system and want to be able to override automatic decisions [31]. To meet these requirements, the *Energy Management Panel* (EMP) has been developed for the interaction between the user and the BEMS [4].

The main view of the *visualization* gives an overview of the most important energy flows in the building (see Fig. 3). The upper section focuses on electric

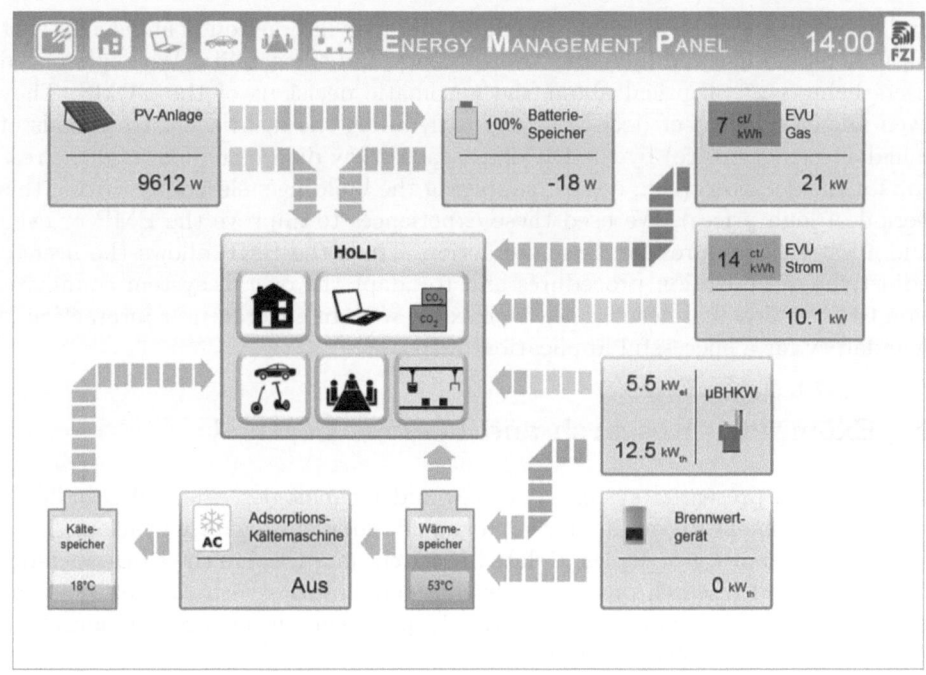

Fig. 3. Energy management panel: visualization of energy flows in the HoLL

energy flows and the lower section highlights thermal energy flows. The user is able to retrieve more detailed information of each component, e.g., energy profile of the PV generation or µCHP unit's operating condition, by selecting relevant detailed views. This overview supports the user to understand the relation between energy supply, consumption, and storage within the building easily.

Additionally, *parametrization* of BEMS is an important feature in the context of user interaction. The EMP enables the user to define certain parameters, which are mandatory for the optimization of the BEMS. The required parametrization ranges from the definition of *degrees of freedom* for load-shifting of appliances to the definition of overall optimization targets, e.g., maximizing the economic benefit or maximizing the usage of renewable energy sources.

In contrast to the parametrization, which is performed at run time of the BEMS, *system configuration* is carried out before the initial start of the BEMS or after major configuration changes. The system configuration enables the adaption of the generic BEMS architecture, components, and drivers to specific requirements and system settings of individual buildings.

Combining visualization, parametrization, and system configuration, the EMP enables the user to get an overview of the energy flows in the building and to understand the mostly automated decisions of the BEMS. Although we found that most users are not willing to control their BEMS manually in the long term, the option for manual interaction with the BEMS increases the user convenience and acceptance significantly [4]. Furthermore, user interaction is able to prevent misunderstandings and increases transparency and traceability. We experienced users being very skeptical about the automatic decisions of the BEMS. They even assumed incorrect decisions (e.g., scheduling the dishwasher to periods of a high electricity price) by the BEMS, because they did not understand the reason for the decisions (e.g., a CHP supplying the building's electricity during this period of high prices). We used these experiences to improve the EMP or even add additional features for user interaction. Thus, the EMP allows the user to adjust the optimization procedures and to adapt the overall system configuration to her individual needs in a very flexible way. In general, user interaction is mandatory for a successful application of BEMS.

6 Exemplary Research and Lessons Learned

Based on the presented systems, devices and technologies, the HoLL builds a sound environment to conduct research and development regarding various *smart building* and *smart grid* topics. Related research conducted in the HoLL includes publicly funded research projects as well as cooperations with companies of the automation, the automotive, the electrical equipment, the energy, the household appliance, and the software industry.

6.1 Exemplary Research

The trigeneration system of the HoLL, consisting of a µCHP unit, an adsorption chiller, and storages tanks for hot and chilled water, has been analyzed, simulated, and optimized in [26]. The results show that an integrated scheduling of the µCHP unit and the adsorption chiller reduces the average total monthly costs on average by about 16 %, in comparison to up to 11 % when scheduling only the µCHP unit respective the adsorption chiller. This supports the approach of an integrated optimization of all devices that are used within a building. In order to enable the successful integration of the multitude of different appliances into one BEMS, the original architecture of the OSH has been enhanced, based on the experience gained in the HoLL [27].

Detailed recorded load profiles of the PV and battery storage system with a resolution of up to one second and corresponding voltage levels have been used in different contexts, such as the optimization of heat pump operation for demand side management [11,24], the optimization of a trigeneration system as depicted in the previous subsection [26,27], the economic evaluation of local PV generation in electric vehicle car parks [35], and the provision of ancillary services by building energy management systems [5].

The EMP is the user interface of the BEMS, i.e., the OSH, in the HoLL. Thus, the BEMS is operating and optimizing the heating and electricity system in combination with several building automation systems and the EMP provides the central user interaction interface of a BEMS in real-life conditions. Since the HoLL is continuously extended by new components and systems, the EMP is also adapted to meet the new requirements, e.g., combining room reservations from the calendar with air conditioning and heating in the meeting room or providing detailed measurement data of the heating and cooling system. Due to the fact that the BEMS is integrated in the real operation, we get feedback from the users who are working every day in the building. Therefore, the EMP is continuously adapted to the available systems and the users' needs.

The *Living Lab smartHome/AmbientAssistedLiving* is used as a development and demonstration platform within various research projects with companies of the automation and the household appliance industry. Conventional and hybrid appliances, i.e., appliances that utilize more than one energy carrier, are equipped with communication modules and integrated into energy management systems [28]. Therefore, suitable interfaces are being developed that enable energy management functionality as well as value-added services in the domains of comfort, safety, and security, based on the insights gained in the HoLL.

6.2   Transfer from Research to Practice

The applied research and development infrastructure, and the continuous interplay of research and practice, allows developing and evaluating innovative smart building solutions. In multiple cooperations with industry partners valuable knowledge regarding the advancement of new technologies are transferred into practice. In one research project with an industry partner, e.g., the energy management concepts developed in and the experiences made with the HoLL are transferred to a large-scale energy management system for a concrete facility. It is composed of about 30 large commercial and industrial buildings with centralized and decentralized energy equipment. The goal is to optimize the interplay of energy supply and demand in order to increase the green energy share and to decrease costs by providing grid-supporting services. In another project, knowledge regarding the control of thermal hardware has been used for the grid-responsive optimization of heat pumps. A predictive control scheme for heat pumps was developed and—in cooperation with an engineering office—realized on real hardware. It allows reacting to demand and supply fluctuations at the energy exchange market. Together with an manufacturer of solar heating solutions and an energy utility, this solution is currently rolled out in up to 20 buildings in order to gain deeper insights. A further project addresses the realization of a market-ready energy management gateway for private as well as for commercial customers. In this context existing equipment, and in particular interdisciplinary knowledge regarding information and communication technologies for the interconnection of various building entities is used for the advancement of real product prototypes.

6.3   Lessons Learned

Due to the large diversity of consumers, suppliers, storage, and automation systems on the market, the standardized and non-proprietary integration of the relevant components into a BEMS is challenging. Although there are several approaches developing communication standards, these are still at their beginning. Therefore, developing the hardware abstraction component of the BEMS operating the HoLL was extensive and complex. Because of the significant interdependencies between thermal and electrical energy flows in a building, optimization strategies of a BEMS have to take different energy carriers into account.

Additionally, research on the needs of users plays an important role regarding user acceptance of, e.g., comfort, safety, and security functions. Research approaches in energy related labs often focus only on one of these aspects independently. Having operated a BEMS and analyzed and controlled numerous components and systems in the HoLL in different conditions during several years, we came to the conclusion that the success of BEMS is enabled by a combination of these features, i.e., standardization, integrated optimization of all energy carriers, and user-centric automatic energy management.

7   Similar Research Environments

Residential as well as commercial building environments are in the focus of the HoLL. Usually, research environments focus either on *smart home* or on *smart building* environments. The term smart home refers to all buildings that are used by private persons, i.e., dwellings, no matter whether it is a single person apartment, a single-, or a multi-family house. In contrast, smart building refers to all buildings that are used commercially, e.g., office buildings, enterprise buildings, and small industrial buildings [37].

Application-oriented and close to the market research in smart home respective smart building environments is done at the two buildings of the *Fraunhofer-inHaus-Zentrum* [13]. Fields of interest are construction systems, living, ambient assisted living, health care, hotel, and office. The buildings are equipped with a geothermal system, a µCHP, phase change material for energy storage, an absorption chiller, and a smart home server with visualization panels. However, the focus lies on technologies targeting at comfort maximization and energy efficient building operation.

A platform for application-oriented research on smart homes with focus on ambient assisted living, energy efficiency, building automation protocols, and *Internet of Things* is provided by the *iHomeLab* [18] of the *Lucerne University of Applied Sciences and Arts*. A personal energy management and visualization infrastructure based on IPv6 is in development in context of the research project *IMPACT*. The energy management focus, however, is on univalent household appliances.

Hands-on experience and research on smart homes in smart microgrids is done at the *Smart Microgrid Lab* [22] of the *Lakeside Labs GmbH*, focusing on smart metering, self-organizing smart appliances, demand response and load scheduling. It is equipped with a PV system, a battery storage system as well as a data acquisition system and can be operated in grid or island mode. Yet, it considers electrical energy and smart homes environments only.

The *SmartHOME* [36] of the German armed forces is a lab environment for research on security, energy efficiency, and comfort in smart homes and smart buildings. It focuses on sensor and actor networks for building automation and the optimization of heating, ventilation, and air-conditioning systems. Integrated multi-energy carrier optimization and intelligent appliances are not considered.

A research and presentation platform for energy management systems is provided by the *IT4Energy Lab* [14] of the *Fraunhofer Institute for Open Communication Systems*. It focuses on building automation, energy management, renewable energy sources, smart metering, and virtual power plants. The project *envyport* aims at the development of an energy management gateway. Nevertheless, the research focuses solely on electricity as energy carrier and lacks an comprehensive view on energy.

Research in the fields of renewable energy sources, energy efficiency, electricity generation, distribution, and storage systems is done at the *ServiceLab Smart Energy* [15] of the *Fraunhofer Institute for Solar Energy Systems*. It aims at products that are close to the market, which require functional, integration, communication, and conformity testing. However, it lacks the integration of heterogeneous building automation systems.

Energy management and building automation in the context of a smart grid is focused on by the *fortiss Smart Energy Living Lab* [12] at the *Technical University Munich*. It consists of a PV and a battery storage system and includes an office environment that is equipped with a control software, which adapts the local energy consumption to the available energy. This lab focuses on electricity and efficient rule-based control strategies.

The *Energy Smart Home Lab* [19] at the KIT is a research and demonstration environment focusing on smart homes as part of a smart grid. It is equipped with a µCHP unit, a PV system, an air conditioner, thermal storage, intelligent appliances and an electric vehicle which are all monitored and controlled by an integrated energy management system. In contrast to the HoLL, the focus is on private households with homogeneous household appliances.

8   Conclusion and Outlook

Currently, energy systems are subject to deep-going changes due to an increasing share of fluctuating renewable energy sources and distributed generation. To cope with these changes, BEMS enable to turn traditionally passive consumers into active grid participants that leverage their local flexibilities according to the grid's needs. In this paper we presented the *FZI House of Living Labs*, a real-world research and demonstration platform. It was shown how it integrates a multitude of different devices and technologies with heterogeneous interfaces and protocols into a single BEMS which enables the flexibilization of demand and supply in the building. This way, a comprehensive research platform is provided for investigating various aspects in the context of smart buildings. This includes in particular the integration of hardware components with heterogeneous protocols and different communication media, as well as software and algorithms for the prediction and efficient optimization of various devices in different scenarios.

Permanent expansions of the HoLL and numerous collaborations with different stakeholders from the domains of smart grids, building automation, energy management, and automotive assure an appropriate up-to-date environment for the development and evaluation of ICT for future buildings in on-going and future projects. In particular, the optimization of the usage of the battery storage system and the hybrid appliances, a more advanced integration of the electric vehicles, as well the extension of optimization algorithms to support additional kinds of demand response will be in focus of future research.

References

1. Abras, S., Ploix, S., Pesty, S., Jacomino, M.: A multi-agent home automation system for power management. In: Cetto, J.A., Ferrier, J.-L., Costa dias Pereira, J.M., Filipe, J. (eds.) Informatics in Control Automation and Robotics, vol. 15, pp. 59–68. Springer, Heidelberg (2008)
2. Allerding, F., Mauser, I., Schmeck, H.: Customizable energy management in smart-buildings using evolutionary algorithms. In: Esparcia-Alcázar, A.I., Mora, A.M. (eds.) EvoApplications 2014. LNCS, vol. 8602, pp. 153–164. Springer, Heidelberg (2014)
3. Allerding, F., Schmeck, H.: Organic smart home: architecture for energy management in intelligent buildings. In: Proceedings of the 2011 Workshop on Organic Computing, pp. 67–76. ACM (2011)
4. Becker, B.: Interaktives Gebäude-Energiemanagement. Ph.D. thesis, Karlsruhe Institute of Technology (2014)
5. Becker, B., Mauser, I., Schmeck, H., Hubschneider, S., Leibfried, T.: Smart grid services provided by building energy management systems. In: Innovative Smart Grid Technologies (ISGT-LA), 2015 IEEE PES, October 2015
6. Bluetooth SIG Inc.: http://www.bluetooth.com/Pages/Bluetooth-Smart.aspx
7. Crawley, D.B., Hand, J.W., Kummert, M., Griffith, B.T.: Contrasting the capabilities of building energy performance simulation programs. Build. Environ. **43**, 661–673 (2008)
8. Dawson-Haggerty, S., Krioukov, A., Taneja, J., Karandikar, S., Fierro, G., Kitaev, N., Culler, D.E.: BOSS: building operating system services. In: NSDI 2013, pp. 443–458 (2013)
9. EEBus Initiative e.V.: http://www.eebus.org
10. EnOcean GmbH: https://www.enocean.com
11. Fassnacht, T., Loesch, M., Wagner, A.: Simulation study of a heuristic predictive optimization scheme for grid-reactive heat pump operation. In: Proceedings of the REHVA Annual Conference 2015. REHVA (2015)
12. fortiss GmbH: fortiss - Smart Energy Living Lab. http://www.fortiss.org/forschung/projekte/smart_energy_living_lab
13. Fraunhofer-Gesellschaft zur Förderung der angewandten Forschung e.V.: Das Fraunhofer-inHaus-Zentrum. http://www.inhaus.fraunhofer.de
14. Fraunhofer Institute for Open Communication Systems: IT4Energy Lab. https://www.fokus.fraunhofer.de/de/it4energy/lab/smart-metering
15. Fraunhofer Institute for Solar Energy Systems: ServiceLab Smart Energy. http://www.ise.fraunhofer.de/de/servicebereiche/servicelab-smart-energy/smartenergylab

16. General Electric: http://uk.geindustrial.com/products/home-automation/
17. Hellfeld, S., Oberweis, A., Wessel, T.: Plattform zur prozessgetriebenen Entwicklung von anwenderinduzierten Innovationen in domänenbergreifenden Anwendungsszenarien. HMD Praxis der Wirtschaftsinformatik **52**(3), 337–346 (2015)
18. Hochschule Luzern: iHomeLab - Forschungszentrum für Gebäudeintelligenz. http://www.ihomelab.ch
19. Karlsruhe Institute of Technology: Energy Smart Home Lab. http://www.izeus.kit.edu/english/57.php
20. Kieback & Peter GmbH & Co. KG: http://www.enkey.de
21. KNX Association cvba: http://www.knx.org
22. Lakeside Labs GmbH: Smart Microgrid Lab. http://www.lakeside-labs.com/about/laboratories/
23. Loesch, M.: TRNSYS-Java-Coupler (TRNSYS Type 299), open source (2013). http://sourceforge.net/projects/trnsys-java-coupler/
24. Loesch, M., Hufnagel, D., Steuer, S., Fassnacht, T., Schmeck, H.: Demand side management in smart buildings by intelligent scheduling of heat pumps. In: Proceedings of the IEEE International Conference on Intelligent Energy and Power Systems (IEPS 2014). IEEE (2014)
25. Mauser, I., Dorscheid, M., Allerding, F., Schmeck, H.: Encodings for evolutionary algorithms in smart buildings with energy management systems. In: 2014 IEEE Congress on Evolutionary Computation (CEC), pp. 2361–2366. IEEE (2014)
26. Mauser, I., Feder, J., Müller, J., Schmeck, H.: Evolutionary optimization of smart buildings with interdependent devices. In: Mora, A.M., Squillero, G. (eds.) EvoApplications 2015. LNCS, vol. 9028, pp. 239–251. Springer, Heidelberg (2015)
27. Mauser, I., Hirsch, C., Kochanneck, S., Schmeck, H.: Organic architecture for energy management and smart grids. In: The 12th IEEE International Conference on Autonomic Computing (ICAC 2015). IEEE (2015)
28. Mauser, I., Schmeck, H., Schaumann, U.: Optimization of hybrid appliances in future households. In: Proceedings of International ETG Congress 2015: Die Energiewende - Blueprint for the New Energy Age. VDE (2015) (in press)
29. Mültin, M., Allerding, F., Schmeck, H.: Integration of electric vehicles in smart homes - an ICT-based solution for V2G scenarios. In: Proceedings of the 2012 IEEE PES Innovative Smart Grid Technolgies Conference. IEEE (2012)
30. Paetz, A.G., Kaschub, T., Jochem, P., Fichtner, W.: Load-shifting potentials in households including electric mobility - a comparison of user behaviour with modelling results. In: 2013 10th International Conference on the European Energy Market (EEM), pp. 1–7, May 2013
31. Paetz, A.G., Becker, B., Fichtner, W., Schmeck, H.: Shifting electricity demand with smart home technologies - an experimental study on user acceptance. In: 30th USAEE/IAEE North American Conference Online Proceedings, vol. 19, p. 20 (2011)
32. Plugwise BV: https://www.plugwise.com
33. RWE Effizienz GmbH: https://www.rwe-smarthome.de
34. Soares, A., Antunes, C.H., Oliveira, C., Gomes, Á.: A multi-objective genetic approach to domestic load scheduling in an energy management system. Energy **77**, 144–152 (2014)
35. Steuer, S., Gärttner, J., Schuller, A., Schmeck, H., Weinhardt, C.: Economic evaluation of local photovoltaic generation in electric vehicle car parks. In: VDE-Kongress 2014 Smart Cities. VDE VERLAG GmbH (2014)
36. Universität der Bundeswehr München: SmartHOME. http://smarthome.unibw-muenchen.de/

37. VDE Verband der Elektrotechnik Elektronik Informationstechnik e.V.: Die deutsche Normungsroadmap Smart Home + Smart Building, online, Frankfurt (2013)
38. Wright, A., Firth, S.: The nature of domestic electricity-loads and effects of time averaging on statistics and on-site generation calculations. Appl. Energy **84**(4), 389–403 (2007)
39. ZigBee Alliance: http://www.zigbee.org

Electric Mobility

Data-Based Assessment of Plug-in Electric Vehicle Driving

Jürgen Wenig[1,2], Mariya Sodenkamp[1(✉)], and Thorsten Staake[1,3]

[1] Energy Efficient Systems Group, University of Bamberg, Bamberg, Germany
{juergen.wenig,mariya.sodenkamp,
thorsten.staake}@uni-bamberg.de
[2] Information Systems Engineering, University of Würzburg,
Würzburg, Germany
[3] Department of Management, Technology and Economy, ETH Zurich,
Zurich, Switzerland

Abstract. The limited driving range of electric vehicles (EV) is one of the biggest deployment challenges for electromobility. We use GPS driving data from a fleet of about 1,000 conventional private vehicles collected over two years to simulate energy consumption of electric cars. We estimate how much energy is required for EV charging at home and at a secondary parking location (e.g., at work) and to what extent energy from solar panels during sunlight hours can be used for charging.

Keywords: Electric vehicle · GPS driving data · Energy demand · Photovoltaics

1 Introduction

Electric vehicles (EVs) are a crucial element of our society's mobility strategy as they provide independence from fossil fuels, can lead to a reduction of CO_2 emissions, and lay the foundation for technology leadership, job creation, and economic growth [1, 2]. From the perspective of the automotive industry and utility companies, the transition toward EVs opens up new challenges and opportunities within the emerging EV value chain. For drivers, electrification seems to be the most viable trend in the long run on the pathway to sustainable petroleum displacement [3].

However, both supply side actors (including EV, battery, and automotive component manufacturers) and demand side actors (drivers and governments) face significant barriers that hamper a rapid mass-market adoption of EVs. The industrial success depends, on the one hand, on the competencies of new business players to meet the demands of the new emerging ecosystems (particularly battery pack producers), and, on the other hand, on the availability of the EV-enabling infrastructure. Both factors ultimately determine range, the ease of re-charging the car, and thus the degree to which the consumers' requirements regarding personal mobility will be satisfied. In fact, the popularity of electric cars in the years to come is likely to depend much more on improvements to their performance and on the success in overcoming the lack of infrastructure than on the oil price [4]. Although the developments to date suggest

© Springer International Publishing Switzerland 2015
S. Gottwalt et al. (Eds.): EI 2015, LNCS 9424, pp. 115–126, 2015.
DOI: 10.1007/978-3-319-25876-8_10

significant potential, the actual global uptake on EV sales is not yet impressive. Main reasons include high acquisition costs, the poor charging infrastructure, and the limited driving range [5, 6]. Moreover, EVs represent a new source of demand for electricity and may require the addition of electricity-generation capacity and substantial upgrades to transmission and distribution infrastructure. So, for instance, [7] predicts that widespread use of EVs will increase electricity demand modestly – in the order of 10–15 %. Reduction of this demand can be achieved by using distributed resources of renewable energy, like solar panels on a rooftop. This is where the motivation of our research emerges.

In this paper, we provide estimates on the two critical factors of electric mobility: range and electric load for two likely-to-come charging topologies (charging at home and at work). Based on the real-world driving data collected from vehicle sensors of conventional cars - specifically, Global Positioning System (GPS) based location data - *we aim at (i) estimating the share of the trips that can be driven fully electrically, (ii) examining the electricity demand change, and (iii) assessing the demand change in the situation when EVs charge from the solar energy during sunlight hours.*

To reach our goals, we assess energy demand of the simulated EVs based on the real-world car motion patterns and examine two charging scenarios: (a) charging at home, and (b) charging at home and at a secondary parking location (e.g., work). Parking locations of each car are found using a density based clustering algorithm. A comprehensive physical car energy consumption model is employed to derive state of charge during the trips. The underlying assumption in our model is the use of range extenders, so that all actual trips are taken into consideration.

Analysis of car motion profiles is an emerging stream of the EV feasibility analysis and infrastructure planning research. The idea is to use high-resolution data from gasoline vehicles collected through GPS and derive individuals' driving patterns. Recently, [8] analyzed start and end locations from the data including GPS coordinates and speed profiles of 79 drivers in Michigan over 5 nonconsecutive days to simulate a plug-in hybrid electric car and provide information on the likely state of charge of the battery at the time of arrival. [9] statistically investigated daily driving distances of 484 car owners in the United States and concluded that even short-range EVs can be satisfactory for a significant fraction of the population. In general, these approaches help to segment vehicle buyers by the range needs. Driving profiles associated with 255 households were analyzed by [10] to address the EV household needs in Seattle for one- and multiple-vehicle dwellings. [11] recorded usage data of 76 cars in a one–year period in the city of Winnipeg, Canada, and used this data in a simulation model to predict EV charging profiles and electrical range reliability. [12] evaluated GPS based travel surveys and found that 10 % of drivers could reach all destinations electrically when charging at home. [13, 14] used GPS driving data from the cities of Modena and Firenze in Italy to estimate the percentage of urban trips that could be covered electrically. We go beyond the state of the art by comparing two realistic charging scenarios, using a larger sample over a longer period of time in Europe that may reveal different driving characteristics and thus produce different results regarding the range and electricity demand. The experimental part of our research uses the data from 985 cars with internal combustion engine in North Italy collected over two years from 2007 to 2009 at a 2-km granularity. Moreover, in contrast to the datasets used in the reported studies, our data cover different kinds of areas (not only urban, but also suburban and rural).

The results indicate that about *63 % of distances could be driven electrically* if charging was possible at home only, given that each driver has a home-base parking spot with a socket. With an additional charging facility (e.g., at work), this number rises only modestly to 69 %. The availability of fast charging facilities (i.e., charging 80 % of an empty battery within 30 min) increases the electrically driven share to 71 %. Furthermore, the analysis of *network load* shows that EV charging intensifies electricity demand peaks in the evening hours (in an unmanaged charging scenario), in particular during the working days (Monday to Friday), when drivers return home to charge their vehicle. The introduction of a secondary charging facility can slightly smooth this demand peak without load shifting strategies. Finally, the comparison of charging energy demand with sunlight hours shows that a direct *use of photovoltaic cells for EV charging* at home may be limited. While cars are parked at home, the sun is shining 44 % of the standing time on average. In contrast, secondary charging facilities are used during the daytime with the sun shining 68 % of the parking time (which is an advantage for mostly flat production buildings that inhibit large spaces for PV modules, but a disadvantage for urban areas with high office buildings that offer less roof area per employee).

The remainder of the paper is organized as follows. Section 2 provides description of the data that supports our calculation. The analysis methodology is described in Sect. 3. Section 4 details the results. Conclusions are summarized in Sect. 5.

2 Data Description

We obtained the experimental data from the major European pay as you drive insurance provider. To collect it, combustion engine cars were equipped with an on-board GPS sensor and a GSM module. During vehicle operation, position updates were carried out every few seconds and then aggregated on the device level to reduce costs of transmission and storage. For aggregation, the system calculated traveled distance from incremental position updates and generated new data entries every 2 km. The distance between the points can in some cases exceed 2 km intervals if no position update was available, for example, due to the signal obstruction. In our study, we focused on the data from accident-free cars. The same dataset and a similar data cleansing approach (except elimination of drivers exhibiting many long trips) was used in [15]. As a result, 985 vehicles were considered. For privacy reasons, no driver particulars of any kind were included in the sample. Table 1 outlines the available variables.

3 Data Analysis Methodology

We present an approach that uses GPS data to simulate driving, energy consumption, and charging of EVs. The proposed procedure includes three following steps:

Table 1. Variables available in the GPS dataset [15]

Attribute	Description
Car ID	Car/device ID
Date and time	Date and time on which the dataset was recorded
Latitude, Longitude	Vehicle position in decimal notation
Speed	Vehicle speed at recording time in km/h
Distance to previous point	Distance traveled since last recording point
Time since previous point	Time traveled since last recording point
Panel session	Provides dataset description/dataset purpose:
	0: Dataset recorded on ignition turn-on
	1: Dataset recorded during vehicle operation
	2: Dataset recorded on ignition turn-off
Road type	Road type at recorded location:
	0: Urban, 1: Highway, 2: Extra Urban

1. *GPS data exploration*. This step converts raw GPS data into driving trajectories. The term 'trajectory' refers to a connected sequence of GPS measurements. These sequences allow for identifying complete trips, which may be described by their length, duration, speed and acceleration. In addition, the parking location and time between successive trips is determined.
2. *Stay region clustering*. Based on the trip information, it is possible to identify potential charging areas, that may include a home base, a workplace, or commercial sites [16]. We identify the hypothetical home location and a second major parking place of each vehicle by means of density based clustering algorithm.
3. *EV simulation*. Finally, we simulate energy consumption of EVs during the trips, along with their charging at home and at the second major parking location. This is necessary for determining which destinations are reachable. The energy consumption model reflects one of the most relevant aspects of electric driving: the state of battery charge, which depends on the battery capacity and on the driving pattern.

3.1   Derive Trips from GPS Data

Start locations and destinations of trips are derived from the panel session data. We convert this information into driving trajectories to identify trips, which are described by their length, duration, and speed. A complete trip is a sequence of connected GPS measurements between a start location and destination.

3.2   Detection of Charging Locations

EVs can be charged at any conventional power outlet. Therefore, charging facilities may be installed at many different locations, for example, at home, at commercial sites, or at work [16]. Based on the trip data, it is possible to identify potential charging areas, such as the home location or a secondary parking location (e.g., workplace of a driver or a secondary residence).

Numerous studies indicate that charging overnight at home is a typical and convenient method [8, 9, 13, 14, 17]. [18, 19] suggest that the first and the last location during the working day may refer to the driver's home, and that the most stable location during the day may be the work site. [20] assumes that the home location is often the last destination of the day, the location where more time is spent than at any other place, or that a driver is at home at a certain time.

Inference of the home location from GPS driving data can be made using heuristic or clustering approaches [18, 20]. The choice of the appropriate method depends on the nature of the data at hand, as well as on the exact application purpose [21]. For instance, [20] uses hierarchical location clustering and assumes that the largest cluster refers to the driver's home. [22–24] suggest partitioning clustering for finding home bases. [18, 25] use density based clustering algorithms, that were verified with survey participants' home addresses in [26]. The second densest cluster was identified as a workplace.

Clustering algorithm for the home base search must be able to deal with noise (i.e., points reflecting occasional/irregular stops that do not belong to any cluster). In addition, the algorithm has to be able to deal with arbitrarily shaped clusters, because the shape of the potential stay regions is not necessarily circular or convex but depends on available parking sites. We use a popular density based clustering algorithm DBSCAN [27] to find dense groups of parking locations. DBSCAN uses two input parameters: First, the parameter *Eps* defines a neighborhood radius of a parking point, while iteratively including points into the cluster. Second, the parameter minimum number of parking events (*MinPts*) within *Eps* is used to define the cluster density. As a result, the method identifies the home base as a cluster with the maximum overall parking duration. We empirically tested different parameters with the assumption that cars park at home over night, so that *MinPts* = 20 was finally chosen, and *Eps* was incrementally approximated for each vehicle by setting the home base area (respectively its convex hull) to 100,000 square meters. Similarly, the secondary parking location (e.g., workplace or another residence) is a cluster with the second longest overall parking duration.

3.3   Electric Vehicle Simulation

We estimate energy consumption of EVs and derive their charging duration based on the movement patterns of conventional cars. We are aware that driving patterns might be different between a conventional car and an EV with range extender, yet we assume that mobility habits of individuals are rather stable. We assume that real-world data on EV driving that currently becomes available would probably be even more biased as early adopters tend to preferably use EVs as second cars and thus that current driving habits are a sufficient approximation of future electric mobility patterns.

Information on the length and the average speed of trip sections is used to estimate energy consumption during the trips based on the realistic energy consumption model for cars [28]. The model reflects one of the most relevant aspects of electric driving: the state of charge (SOC) of the battery, which depends on the battery capacity and the driving pattern from which the energy consumption of the vehicle is derived. Hereby, rolling resistance [29–31], aerodynamic drag [28, 29, 32], and acceleration [30] are taken into account. With high granularity of the driving data, acceleration can be derived from the speed. However, due to the distance of about 2 km between the pairs of consecutive GPS measurements in our data, such an approach would be rather inaccurate. A comparison of driving cycles [33] indicates that especially in urban areas cars accelerate and decelerate more frequently. Therefore, we use driving cycles (i.e., series of speed points versus time) for the estimation of acceleration and deceleration by replicating speed profiles from a database of driving cycles [34] that was built within the ARTEMIS project [35]. The ECE-15 and EUDC driving cycles refer to the urban and extra urban driving patterns respectively. The EMPA T130 is used for highway driving. We assume that with a negative total tractive effort caused by deceleration, 50 % of energy could be recuperated when braking [28].

The amount of charged energy depends on the state of charge, on the available charging power, and on the time a vehicle parks at the charging location. Only parking locations with a minimal parking duration of 15 min were considered for charging in our study. According to [36], 80 % of the of the total required charge will be provided in a linear way while the remaining 20 % require thrice the charging time.

4   Results

We compare scenarios where EVs are charged (a) only at home, and (b) at home and at a secondary parking location. We use parameters for our energy consumption model that roughly resembles a BMW i3. For our GPS driving dataset, we calculate an average energy consumption of 14.3 kWh per 100 km by applying the model from [28]. This slightly exceeds the official energy consumption of 12.8 kWh per 100 km according to the manufacturer, most probably due to the optimistic consumption statements that are based on the ECE test cycle [30]. For charging the simulated EV batteries, we assume a charging power of 3.7 kW (16 A) [30].

4.1   Shares of Electrically Drivable Distances and Gasoline Savings

Scenario 1: Charging at home. We found that *77 % of destinations could be reached electrically if the EVs are only charged at the estimated home charging facilities*. This corresponds to 63 % of mileage that could be driven electrically. In Fig. 1a) we provide a distribution of electrically driving shares, which indicates how many cars could reach what portion of mileage electrically. The remainder could be reached by means of a range extender, e.g., the commercially available one for the BMW i3 [30].

Our calculations indicate that the average driver could save 669 l or 59 % of required gasoline per year, if an electric engine was used instead of an internal combustion engine. Differences between electrically drivable mileage and gasoline savings indicate that variations in driving style (in particular long distance trips with higher speed) have to be taken into account when quantifying electric driving potential.

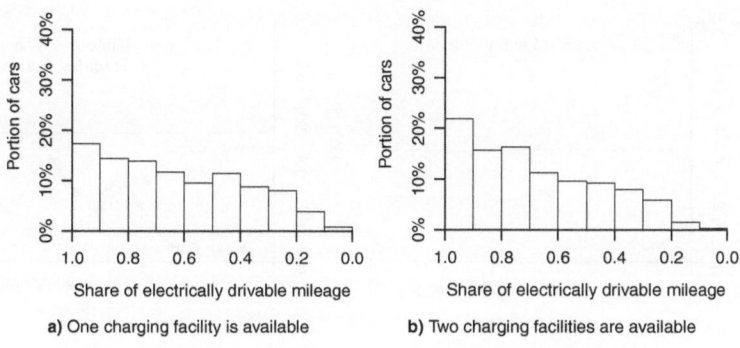

a) One charging facility is available   **b)** Two charging facilities are available

Fig. 1. Share of electrically drivable mileage

Scenario 2: Charging at home and at a secondary charging facility. In this case, *the percentage of reachable destinations rises to 83 %* and the share of electrically drivable mileage rises to 69 %. Figure 1(b) shows the corresponding distribution. The average driver could now save 741 liters or 64 % of required gasoline per year.

4.2   Grid Impact of Electric Vehicle Charging Demand

Recent studies have shown that future electric mobility can have a significant impact on the electric grid. With greater EV shares, the electricity demand grows. However, the main challenge in providing energy for electric mobility lies in increasing peak demand [37]. In particular, in the home charging scenario, demand for vehicle charging is typically higher during peak times in the evening [13, 14]. However, additional charging facilities at work can have a smoothening effect [11]. In this section, we calculate the grid impact of two considered charging scenarios.

Scenario 1: Charging at home. In our simulation, the average vehicle requires 5.3 kWh of energy per day when charging at home. In Fig. 2(a) we show how much energy is required at what time of the day. We can see the peaks at noon and in the evening, when people arrive home. In Fig. 2(b) we compare the weekdays and weekends and see that in general energy demand is lower during the weekend. In particular, demand peaks in the evening are considerably reduced.

Our estimates yield an *increase of electricity demand by 49 % in the EU domestic sector.* To calculate this, we used the average electricity consumption per household and year, which was 3'928 kWh in 2013 according to [38]. Divided by 365 (the number of days in year), it is 10.8 kWh per day. According to the report of the European Commission [39], the residential sector consumes 29.7 % of final electricity. Thus, the *overall increase of electricity demand would be about 15 %.*

Scenario 2: Charging at home and at a secondary charging facility. Average energy requirements for charging per car and day rise to 5.8 kWh in this case, including 4.6 kWh demand at the household side and 1.2 kWh at the secondary location. The EV *grid impact at home is decreased by 13 %* compared to Scenario 1. In Fig. 3(a) we

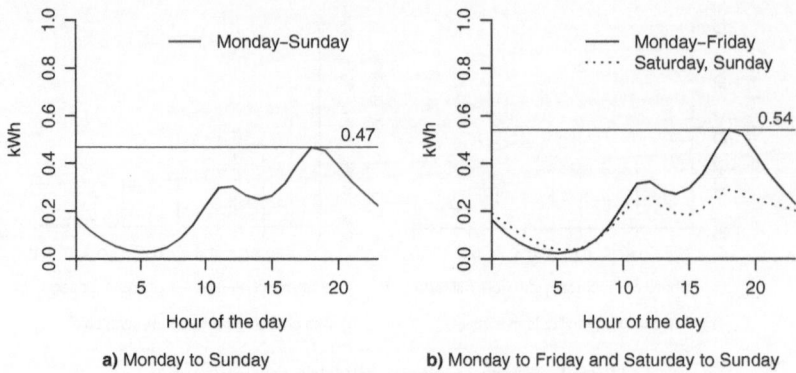

Fig. 2. Average energy needs when charging at home during the day

compare the daily energy demand at home and at the secondary charging facility. At the secondary facility, the peak energy demand is between 7 and 8 AM. The availability of a secondary charging facility also slightly reduces the peak charging demand in the evening. Figure 3(b) shows that daily energy demand is lower during the weekends and that the peak demand in the evening is smoother.

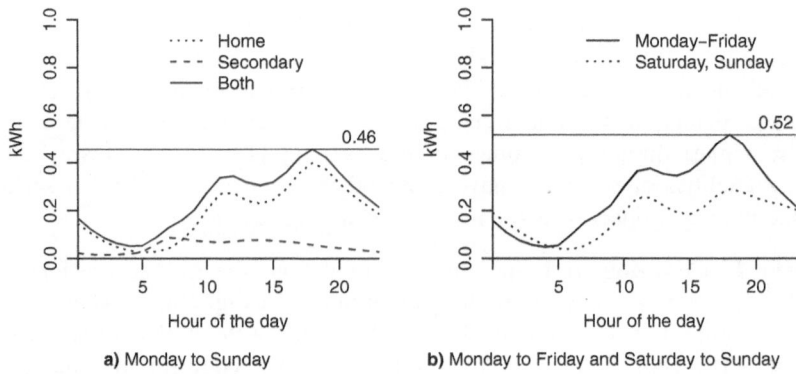

Fig. 3. Average energy needs when charging at home and at a secondary parking location during the day

4.3 Comparison of Charging Times and Sunlight Hours for Electric Vehicle Charging with Photovoltaic Panels

The introduction of EVs increases the utilization of local power generation from photovoltaic panels [40, 41]. Still, the increase is limited because there is a small correlation between photovoltaic power production patterns and plug-in hybrid electric vehicle charging patterns [42]. Analysis of GPS driving data enables us to provide an assessment of the potential of using energy from photovoltaic cells at home for EV charging without temporarily storing electricity. We use hourly historic irradiance data for the considered geographic region and time period, as provided by the MACC-RAD service [43] to determine average sunlight hours at the home bases.

Scenario 1: Charging at home. Our simulation shows that *about 47 % (44 %) of the time EVs charge (park) at home at sunlight.*

Scenario 2: Charging at home and at a secondary charging facility. About 72 % *(68 %) of the time cars charge (park) at the secondary parking location at sunlight,* while 48 % (44 %) of the time they charge (park) at home at sunlight. This leads to the conclusion that solar panels can be particularly useful at secondary charging locations where drivers park during the daytime. The use of photovoltaic energy at home is also promising. However, the use of solar energy increases the importance of EV charging strategies that control flexible loads and keep the electric grid stable [44].

4.4 Impact of Charging Power on EV Reachability and Daily Energy Demand

In our assessment, we use the typical charging power values for the European EVs, which are 3.7 kW (16 A), 7.4 kW (32 A), and 50 kW (125 A) [30, 45]. As can be seen from Table 2, faster charging increases not only destinations reachability and electrically drivable shares, but also daily energy demand and evening demand peaks.

Table 2. Comparison of charging times for the average driver

Charging site	Home			Home and Secondary		
Power	50 kW	7.4 kW	3.7 kW	50 kW	7.4 kW	3.7 kW
Reachability	79 %	77 %	77 %	85 %	83 %	83 %
Driving Share	65 %	64 %	63 %	71 %	69 %	69 %
Daily Energy	5.6 kWh	5.4 kWh	5.3 kWh	6.3 kWh	6.0 kWh	5.8 kWh
Demand Peak	0.58 kWh	0.54 kWh	0.47 kWh	0.53 kWh	0.50 kWh	0.46 kWh
(Time)	(5–6 PM)	(6–7 PM)	(6–7 PM)	(5–6 PM)	(6–7 PM)	(6–7 PM)

5 Conclusion

In this paper, we use real-world GPS driving data from conventional vehicles, a density based clustering algorithm, and an energy consumption model to estimate the potential of electric driving. Our study shows that drivers who would replace their combustion engine car by an EV with an optional range extender could reach about 77 % to 85 % of destinations fully electrically. They could also cover about 63 % to 71 % of their mileage electrically, depending on the availability of charging facilities at home and at a secondary charging facility, and depending on the time required to charge the battery. By comparing parking and charging times with sunlight hours, we observe that there is sunlight about 44 % of the time when cars are parking at home and about 68 % of the time when cars are parking at their secondary charging facility. These results have some direct implications for future assessment of electric mobility, including the assessment of EV battery pack parameter requirements, charging facility parameter requirements, as well as the potential of using privately owned photovoltaic systems for EV charging.

Acknowledgements. The authors would like to thank Prof. Dr. Frédéric Thiesse and the anonymous reviewers for their helpful comments and suggestions.

References

1. European Commission: European Green Vehicles Initiative PPP: Use of new energies in road transport (2013). http://ec.europa.eu/research/press/2013/pdf/ppp/egvi_factsheet.pdf. Accessed 09 April 2015
2. International Energy Agency: Technology Roadmap: Electric and plug-in hybrid electric vehicles (2011). http://www.iea.org/publications/freepublications/publication/EV_PHEV_Roadmap.pdf. Accessed 09 April 2015
3. Ruan, J., Walker, P., Zhu, B.: Experimental verification of regenerative braking energy recovery system based on electric vehicle equipped with 2-speed DCT. In: 7th IET International Conference on Power Electronics, Machines and Drives (PEMD 2014), pp. 1–8 (2014)
4. The Economist: Why the low oil price will not harm sales of electric cars (2015). http://www.economist.com/blogs/economist-explains/2015/02/economist-explains-21. Accessed 09 April 2015
5. Thiel, C., Alemanno, A., Scarcella, G., et al.: Attitude of European car drivers towards electric vehicles: a survey. JRC report (2012)
6. Jean, P.: European Policy on Electric Vehicles (2011). http://ec.europa.eu/enterprise/sectors/automotive/files/pagesbackground/competitiveness/speech-jean-sofia_en.pdf. Accessed 13 June 2015
7. World Nuclear Association: Electricity and Cars: Electric Vehicles (2015). http://www.world-nuclear.org/info/Non-Power-Nuclear-Applications/Transport/Electricity-and-Cars/. Accessed 13 Jun 2015
8. Adornato, B., Patil, R., Filipi, Z., et al.: Characterizing naturalistic driving patterns for Plug-in Hybrid Electric Vehicle analysis. In: Vehicle Power and Propulsion Conference, VPPC 2009, pp. 655–660. IEEE (2009)
9. Pearre, N.S., Kempton, W., Guensler, R.L., et al.: Electric vehicles: how much range is required for a day's driving? Trans. Res. Part C: Emerg. Technol. **19**(6), 1171–1184 (2011)
10. Khan, M., Kockelman, K.M.: Predicting the market potential of plug-in electric vehicles using multiday GPS data. Energy Policy **46**, 225–233 (2012)
11. Ashtari, A., Bibeau, E., Shahidinejad, S., et al.: PEV charging profile prediction and analysis based on vehicle usage data. IEEE Trans. Smart Grid **3**(1), 341–350 (2012)
12. Dong, J., Liu, C., Lin, Z.: Charging infrastructure planning for promoting battery electric vehicles: an activity-based approach using multiday travel data. Trans. Res. Part C: Emerg. Technol. **38**, 44–55 (2014)
13. de Gennaro, M., Paffumi, E., Scholz, H., et al.: GIS-driven analysis of e-mobility in urban areas: an evaluation of the impact on the electric energy grid. Appl. Energy **124**, 94–116 (2014)
14. Paffumi, E., de Gennaro, M., Martini, G., et al.: Assessment of the potential of electric vehicles and charging strategies to meet urban mobility requirements. Transportmetrica A: Trans. Sci. **11**(1), 1–39 (2014)

15. Ippisch, T.: Telematics data in motor insurance: creating value by understanding the impact of accidents on vehicle use. University of St, Gallen (2010)
16. Dickerman, L., Harrison, J.: A new car, a new grid. IEEE Power Energy Mag. **8**(2), 55–61 (2010)
17. Pasaoglu, G., Fiorello, D., Martino, A., et al.: Travel patterns and the potential use of electric cars-results from a direct survey in six European countries. Technol. Forecast. Soc. Change **87**, 51–59 (2014)
18. Gambs S, Killijian M, del Prado Cortez, M.N.: Show me how you move and I will tell you who you are. In: Proceedings of the 3rd ACM SIGSPATIAL International Workshop on Security and Privacy in GIS and LBS, pp. 34–41 (2010)
19. Gambs, S., Killijian, M., Núnez del Prado Cortez, M.: Gepeto: a geoprivacy-enhancing toolkit. In: 24th International Conference on Advanced Information Networking and Applications Workshops (WAINA), pp. 1071–1076. IEEE (2010)
20. Krumm, J.: Inference attacks on location tracks. In: LaMarca, A., Langheinrich, M., Truong, K.N. (eds.) PERVASIVE 2007. LNCS, vol. 4480, pp. 127–143. Springer, Heidelberg (2007)
21. Han, J., Lee, J., Kamber, M.: An overview of clustering methods in geographic data analysis. In: Miller, H.J., Han, J. (eds.) Geographic data mining and knowledge discovery, 2nd edn, pp. 149–187. Taylor and Francis, London (2009)
22. Hoh, B., Gruteser, M., Xiong, H., et al.: Enhancing security and privacy in traffic-monitoring systems. Pervasive Comput. IEEE **5**(4), 38–46 (2006)
23. Freudiger, J., Shokri, R., Hubaux, J.-P.: Evaluating the privacy risk of location-based services. In: Danezis, G. (ed.). LNCS, vol. 7035, pp. 31–46. Springer, Heidelberg (2012)
24. Csáji, B.C., Browet, A., Traag, V.A., et al.: Exploring the mobility of mobile phone users. Physica A **392**(6), 1459–1473 (2013)
25. Zheng, Y., Zhang, L., Xie, X., et al.: Mining interesting locations and travel sequences from GPS trajectories. In: Proceedings of the 18th International Conference on World Wide Web, pp. 791–800 (2009)
26. Smith, R., Shahidinejad, S., Blair, D., et al.: Characterization of urban commuter driving profiles to optimize battery size in light-duty plug-in electric vehicles. Trans. Res. Part D: Trans. Environ. **16**(3), 218–224 (2011)
27. Ester, M., Kriegel, H., Sander, J., et al.: A density-based algorithm for discovering clusters in large spatial databases with noise. KDD **96**, 226–231 (1996)
28. Larminie, J., Lowry, J.: Electric Vehicle Technology Explained. John Wiley & Sons Ltd, West Sussex, England (2003)
29. De Haan, P., Zah, R., Althaus, H.-J., et al.: Chancen und Risiken der Elektromobilität in der Schweiz. VDF Hochschulverlag AG an der ETH Zürich (2013)
30. BMW: BMW i3: Technical data (2013). http://www.bmw.com/com/en/newvehicles/i/i3/2013/showroom/technical_data.html. Accessed 27 April 2015
31. National Institute of Standards and Technology: CODATA Value: standard acceleration of gravity (2014). http://physics.nist.gov/cgi-bin/cuu/Value?gn. Accessed 03 August 2014
32. BMW: The new BMW i3: Technical specification (2014). https://www.press.bmwgroup.com/united-kingdom/pressDetail.html?title=the-new-bmw-i3-press-pack&outputChannelId=8&id=T0154004EN_GB&left_menu_item=node__6728. Accessed 12 August 2015
33. Barlow, T.J., Latham, S., McCrae, I.S., et al.: A reference book of driving cycles for use in the measurement of road vehicle emissions. TRL (2009)
34. André, M.: Database of driving cycles (2006). http://www.inrets.fr/fileadmin/ur/lte/artemis/road3/method31/All_Cycles_in_Artemis_BD_092006.xls

35. André, M.: ARTEMIS 2000–2004 (2009). http://www.inrets.fr/linstitut/unites-de-recherche-unites-de-service/lte/themes-de-recherche/energie-et-pollution-de-lair/parcs-de-vehicules/artemis-2000-2004.html. Accessed 06 February 2015
36. Hsieh, G., Chen, L., Huang, K.: Fuzzy-controlled Li-ion battery charge system with active state-of-charge controller. IEEE Trans. Industr. Electron. **48**(3), 585–593 (2001)
37. Salah, F., Ilg, J.P., Flath, C.M., et al.: Impact of electric vehicles on distribution substations: a swiss case study. Appl. Energy **137**, 88–96 (2015)
38. World Energy Council: Electricity use per household: Electricity Consumption Efficiency (2015). http://www.wec-indicators.enerdata.eu/household-electricity-use.html. Accessed 13 June 2015
39. Bertoldi, P., Hirl, B., Labanca, N.: Energy Efficiency Status Report 2012: Electricity Consumption and Efficiency Trends in the EU-27 (2012). http://iet.jrc.ec.europa.eu/energyefficiency/sites/energyefficiency/files/energy-efficiency-status-report-2012.pdf. Accessed 13 June 2015
40. Salah, F., Flath, C.M.: Deadline differentiated pricing in practice: marketing EV charging in car parks. Comput. Sci.-Res. Dev. 1–8 (2014)
41. Yoshimi, K., Osawa, M., Yamashita, D., et al.: Practical storage and utilization of household photovoltaic energy by electric vehicle battery. In: IEEE PES Innovative Smart Grid Technologies (ISGT), pp. 1–8 (2012)
42. Munkhammar, J., Grahn, P., Widén, J.: Quantifying self-consumption of on-site photovoltaic power generation in households with electric vehicle home charging. Sol. Energy **97**, 208–216 (2013)
43. MINES ParisTech, Transvalor Dpt SoDa: MACC-RAD (2015). http://www.soda-pro.com/web-services/radiation/macc-rad. Accessed 04 June 2015
44. del Razo, V., Goebel, C., Jacobsen, H.: On the effects of signal design in electric vehicle charging using vehicle-originating-signals. Comput. Sci.-Res. Dev. 1–8 (2014)
45. BMW: BMW i3: Charging (2013). http://www.bmw.com/com/en/newvehicles/i/i3/2013/showroom/charging.html#wallbox. Accessed 29 July 2015

Communication and Security

Investigating Wind Farm Control Over Different Communication Network Technologies

Jacob Theilgaard Madsen[1]([✉]), Mislav Findrik[2], Domagoj Drenjanac[2], and Hans-Peter Schwefel[1,2]

[1] Aalborg University, Aalborg, Denmark
{jam,hps}@es.aau.dk
[2] Telecommunications Research Center Vienna (FTW), Vienna, Austria
{findrik,drenjanac,schwefel}@ftw.at

Abstract. Wind energy is one of the main contributors to renewable generation. In order to embed wind farms in Smart Grid concepts, wind turbine controllers have the objective to follow an accumulated total power generation reference while at the same time the controllers aim to reduce the damage (wear out), which the individual wind turbine has to sustain. This paper looks at a centralized control architecture for control of wind turbines, in which sensors at the wind turbine periodically provide their values via a local sensor network and an IP-based wide area network, based on which the controller calculates new set-points. Subsequently, the set-points are sent back to the wind turbine via the IP-based network. Testbed measurements of delays and message loss of different network technologies 2G, 3G, PLC and WLAN are captured while mimicking the control scenario of the wind farm. These measurement traces are fed back into a co-simulation framework to then show the impact on control performance of the different technologies. The results show that 2G and narrow-band PLC cannot support the presented control scenario, mainly due to throughput and delay limitations, while 3G and WLAN technologies are able to provide the necessary communication bandwidth and low delays. The measured delay distributions of the latter two technologies can be used to optimize the scheduling of sensor readings and the benefit from such optimizations is qualitatively discussed.

1 Introduction

Over the last few years we have seen an increase in deployment of renewable energy generation. Due to increased deployment and the fluctuating nature of renewable power generation, there is a greater need for controlling the balance between energy consumption and production, by means of Smart Grid deployments. One large contributor to renewable generation are wind farms. Controlling a wind farm can be done from a central controller which harmonizes the wind farm's power generation by sending set-points to each wind turbine, thus a control is performed via a communication network. This requires a central controller to ensure that each wind turbine produces the required amount of power.

© Springer International Publishing Switzerland 2015
S. Gottwalt et al. (Eds.): EI 2015, LNCS 9424, pp. 129–140, 2015.
DOI: 10.1007/978-3-319-25876-8_11

For this communication network it is possible to use different technologies, however, each technology will behave differently, and thus it is important to analyse the impact of these communication technologies on the controller's performance.

In this paper, we consider a central wind farm controller gathering information from wind turbines distributed over an area. We take measurements of different communication technologies, where we measure packet loss and delays, and use these measurements when simulating the wind farm controller. The comparison of the simulation results of control performance metrics provides a quantitative basis for the suitability of communication technologies for this distributed control scenario.

The type of problem that is described in this paper is within the field of control and automation known as a networked control problem. Reference [1] details the impact of packet loss in a control scenario and attempts to make adaptations to the specific controller. Reference [2] attempts to find maximum allowed delays that still ensures stability of the controller.

Attempts have been made to create testbeds that integrate Smart Grid assets and communication technologies. One such is example is Reference [3] which demonstrates a testbed that includes wired and wireless technologies. It is based on Ethernet and ZigBee networks. Reference [4] introduces a cognitive radio network in a Smart Grid testbed (specifically a microgrid), with the main focus on the security of the Smart Grid. In Reference [5] Zigbee is used as the communication technology, and is implemented in an office environment with the goal to optimize the energy costs for the offices connected to the Smart Grid. Reference [6] details different communication technologies available for wind farm communication, among these are WLAN and ethernet based solutions. Another technology used is fiber optics, as described for a wind farm in Mongolia in Reference [7]. Unlike previous work our methodology is based on measuring different communication network technologies in a scaled down manner compared to the real environment. Subsequently, we use the measurement results by plugging the traces into a co-simulation framework which is detailed in Reference [1].

The rest of the paper is structured as follows: In Sect. 2 we discuss the scenario we are analysing and simulating. Section 3 describes the measurements of different communication technologies, used for the simulations. The results for control performance from a co-simulation model using the measured communication traces are shown in Sect. 4, and we conclude the paper in Sect. 5.

2   Scenario Description

This paper considers a system consisting of a central controller, communicating with N wind turbines over a communication network, each wind turbine containing K sensors. There is a bi-directional information flow, consisting of sensor measurements from the sensors to the central controller, and set points from the controller to the wind turbine. The architecture of the system is shown in Fig. 1 and explained further in Sect. 2.1.

The central controller has the objective to follow a set-point for the wind farms power generation while reducing the fatigue a wind turbine experiences in

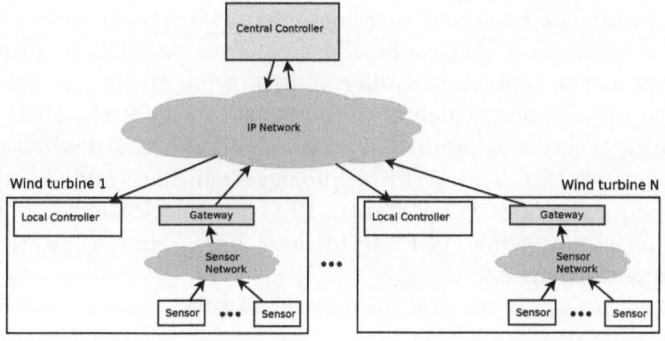

Fig. 1. Overview of the communication network.

order to prolong the lifetime of the wind turbine. The central controller should ensure that each wind turbine in the farm produces a certain level of power to ensure that the entire wind farm reaches the given set-point for the wind farm, motivating the central control architecture of Fig. 1. The sensors send information regarding the state of the wind turbine to the central controller. The central controller uses the sensor values and its internal state to compute set points for the wind turbine. These set points are then forwarded to the wind turbine, where the local controller uses them to adjust the wind turbine. An overview of this scenario can be seen in Fig. 1, where the grayed boxes denote places that introduce delay in the information flow, see Sect. 4.2 for the assumptions taken relating to this scenario.

2.1 Communication Network Architecture

The end-to-end communication path between a sensor and the central controller consists of two communication networks, which can be seen in Fig. 1, with conceptually different properties; a sensor network and an IP network. The sensor network is a communication network internally on the wind turbine, and can be implemented by different communication technologies, for example a fieldbus. It handles the communication between the sensors on the wind turbine, and contains a gateway that handles communication from the wind turbine to the central controller. The IP network is the network between a wind turbine and the central controller. In this paper we will investigate different technologies that can be used to implement this IP-based communication.

In this paper we investigate the effects on the controller performance using four different technologies for the communication network. The technologies are 2G, 3G, WLAN and PLC. These technologies do not require additional hardwired communication infrastructure to be deployed, which can speed up deployment and reduce costs of wind farms. Cellular is already deployed in many regions and hence can be used without further infrastructure. Whether 2G or 3G fulfils the requirements of this scenario is part of the analysis. WLAN is an off-the-self technology readily available and covering a few hundreds of meters which could support wind farms in a geographically smaller area. PLC in principal could be interesting, as the central controller and the wind turbine controller may be connected via power lines, which would mean that except the PLC modems no additional infrastructure is needed. However, unlike the hard-wired communication infrastructure, PLC and WiFi technologies can have additional drawbacks impacting their reliability. For instance, PLC communication is not possible if the power line is cut, while WiFi technology has a shared spectrum possible reducing the performance.

2.2 Controller Design

The controller aims to reduce the fatigue load of the shaft in the wind turbine. It does this while tracking a desired power reference. To achieve this the controller requires an input from a fatigue estimator. This estimator requires sensor measurements of the wind turbine, which are the physical measurements of torque, as well as the angle rotation of the wind turbine. These measurements are run through different hysteresis relays with different weights. For details on estimator synthesis and the controller see [8].

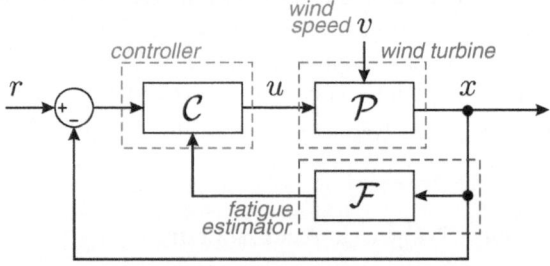

Fig. 2. Centralized Controller scheme without network effects.

The control scheme without network effects is depicted in Fig. 2. This controller describes the procedures for an individual wind turbine, while the same procedure, with coordination on the generation share of each wind turbine, is applied for the whole farm, which requires a centralized controller. For control performance, we later analyse a single wind turbine, while the communication scenario is shaped by the whole wind farm.

2.3 Scenario Specific Parameters

Figure 3 depicts a message sequence diagram of a control period with message delays. The figure shows that during a given control period the controller will calculate a set point for the wind turbine. The set point will be sent to the wind turbine, which will then act upon it. At some point during the control period, defined by the value T_o, denoting the offset time prior to control executions C_i, the sensor will take a reading, sending it to the controller.

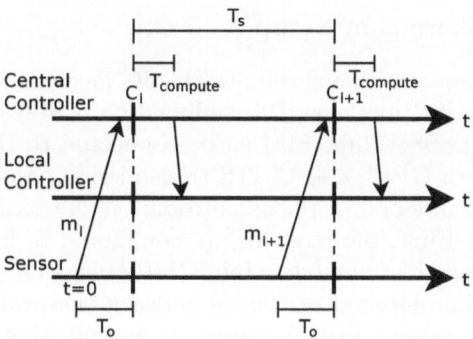

Fig. 3. Message sequence diagram between the sensor and controller for a single wind turbine

The controller cycle, T_s, is determined by the design of the controller, and is in this scenario set to 150 ms with 50 ms controller computation delay, and cannot easily be changed. We consider a total of 30 wind turbines, each with 3 sensors, in the wind farm, which constitutes a small to medium sized wind farm.

As previously mentioned the communication technologies used are 2G, 3G, WLAN and PLC. While these technologies allow communication they do have some limitations. For 2G and 3G there are requirements for each wind turbine to have a sim card. Furthermore, unless a dedicated cell is setup for the wind farm, the wind turbines will have to share the communication medium, and the base station may be placed far from the wind farm thus reducing the quality of service. For WLAN there may be problems with establishing a connection over an entire wind farm, due to the size of the area. This may be alleviated by multi-hop communication with wind turbines relaying end-2-end messages from/to the central controller, however, this will increase the load on each WLAN access point which can potentially lead to additional queueing delays. Furthermore both WLAN and 2G/3G are wireless technologies, thus each wind turbine would share the communication medium with each other, and potentially other users in the area. While PLC does not have the problem of sharing the communication medium with other users in the area, it does have it own problems.

3 Measurements of Communication Technologies

The aim of the measurement setup was to mimic the communication pattern generated between the central controller and 30 gateways on the wind farm communication network in order to characterize the network properties in terms of delays and packet losses. For each of the technologies ICMP ping measurements are performed with 64 bytes of application layer payload in the uplink (i.e. for the sensor measurements) and downlink (i.e. for the control commands) roughly corresponding to an actual size of the messages [9]. In further subsections, details of measurement setups for each of the technologies are given.

3.1   2G/3G Measurement Setup

The test-bed setup for conducting cellular 2G/3G measurements is comprised of three Huawei E392 USB modems [10] equipped with SIM cards from Austrian telecom provider YESSS. Three SIM cards connected to the modems are provisioned to allow both GPRS and UMTS transmission access. Furthermore, the equipment is placed indoors, in the same room on the second floor of an eight floor office building. First, the modems are configured to have access only to a GPRS network. Through each modem ten ICMP ping connections were executed in parallel with a time interval of 150 ms between the probes, thus 30 connections in total were running in order to mimic communication pattern between the gateway and the central controller. Subsequently, the same procedure is repeated for the 3G network by configuring the modems for UMTS transmissions. In both cases, around 4000 measurements are collected from each ICMP ping connection, during a week day between 11 am and 12 am. The resulting distribution of the measurements for 2G and 3G respectively can be seen in Figs. 7 and 4. The distance between the modems and GPRS/UMTS base transceiver stations (BTS) is around 30 m, since the stations are located on top of the same building.

3.2   PLC Measurement Setup

Powerline communication technology is evaluated with narrow-band Devolo G3-PLC 500 k PLC modems [11], having a total throughput capacity of 240 kbps. Five such PLC modems are connected via power-lines of approximately 1 m length in total, which is much shorter than an actual implementation where they would potentially be hundreds of meters long, making this a best case scenario. In an initial test, three parallel ICMP ping connections are ran from three different PLC modems (9 connection on the PLC network), with 150 ms time interval. It is shown later that for nine connections running over a PLC network, the network performance is impractical for the wind farm controller due to high delays and packet losses even in this best case setup with short distances.

3.3   WiFi Measurement Setup

WiFi technologies are tested using two commercially available embedded boards "ALIX 3D2" [12], which allow different IEEE 802.11 hardware modules to be added onto the system. For this study, we connected the 802.11 g (2.4 GHz band) modules of type CM9-GP Wistron [13] onto the boards. The WiFi points are placed indoors within the same room, on the distance of approximately 3 m, line-of-sight, which again is shorter than an actual implementation, making this a best case setup. In the vicinity of the WiFi points 23 other networks were scanned, 16 of them were running on 2.4 GHz band, thus adding some interference to the signal; the transmission power of is set to 15 dBm. Over such WiFi link, 30 parallel ICMP ping connections were executed during a work day between 3 pm and 5 pm, with the same interval as before. The resulting delay distribution observed on the link is shown in Fig. 5.

3.4   Measurements Results

The results of the measurements are used in Sect. 4 as a time series in the co-simulation framework in order to determine the delays of individual messages. The measurements we have access to are, as explained above, RTT ping messages. The delays we are interested in are, however, one-way delays. Therefore we assume that the uplink and downlink channel properties are identical and we can thus divide the RTTs by 2 to get one-way delays. This represents an approximation as many technologies differentiate in their handling of uplink and downlink data streams. All delays shown in this section will be the one-way delays unless otherwise specified.

As can be seen in Fig. 4 there are two peaks to the 3G measurements. The first peak is at roughly 15 ms, and the other, smaller, peak is at 20–22 ms. We have outliers of the data as high as 98.5 ms, which can be caused by spikes in traffic on the base station. In contrast to the measured delay distribution for 3G, the delays of WLAN in Fig. 5 by shape appear like an exponential distribution.

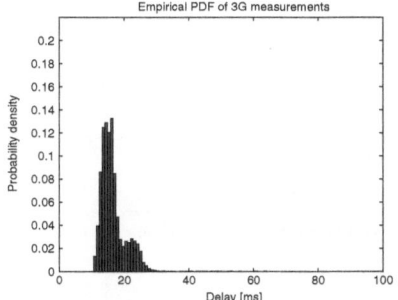

Fig. 4. PDF of the 3G measurement results

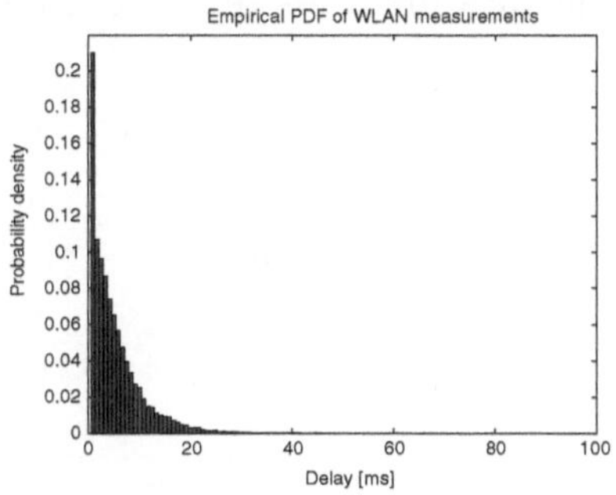

Fig. 5. PDF of the WLAN measurement results

From the 30 2G connections we take the data from a single connection on sim-card 2 and plot it as a time series in Fig. 6. The measurements of the other connections show the same trend as we see in Fig. 6. We see low delays at start of the measurements, which may be caused by not all sim cards and/or connections being setup right away. We see a few spikes in the delays between 0 and up until sample number 4000, which can be caused by spikes in traffic on the base station. After measurement 4000 we start to see a steady increase in delays, most likely caused by increased traffic leading to longer queues on the base station and in the sim-cards. For the evaluation later, we only use the samples from 250 to 4250, in order to avoid the large instability period that occurs after. The subsequent table and Fig. 7 show the distribution of these 4000 samples.

Results from the PLC measurements show that the average delay is several control cycles long, and the packet loss is 78 %, even for only 9 parallel connections. This is rather problematic for the controller, as it will lead to instability and the controller being unable to generate set-points. For this reason these

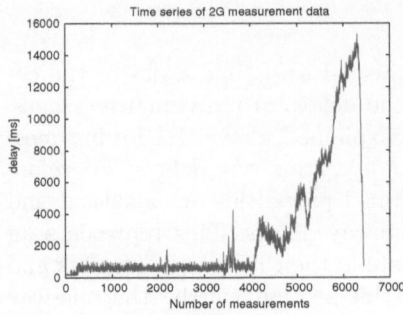
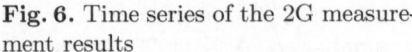

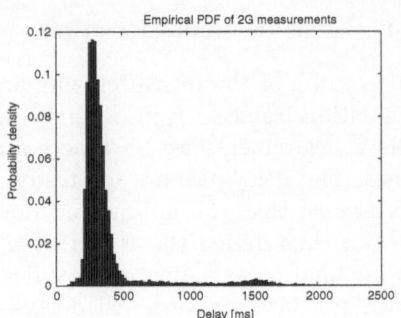

Fig. 6. Time series of the 2G measurement results

Fig. 7. PDF of the 2G measurement results

results are not simulated together with the other performance measurements, as it is clear from the statistics that the solution is not feasible. The statistics for all the measurements results can be seen in the table below.

Technology	Minimum delay	Average delay	Maximum delay	Packet loss
3G	10.5 ms	16.7 ms	98.5 ms	0 %
WLAN	0.5 ms	5.4 ms	83.2 ms	0 %
2G	84.5 ms	385.2 ms	2131 ms	0 %
PLC	50.4 ms	1504 ms	3400.8 ms	78 %

4   Controller Performance Results

In this section we show the results from simulation runs to determine the impact the different communication technologies have on the performance of the controller.

4.1   Co-simulation Framework

A co-simulation framework has been developed, in which MATLAB is used to simulate the wind turbine controller, and OMNeT++ is used as a discrete-event simulator of network delays. The OMNeT++ simulation is using the traces from the performance measurements to impose delays on the messages exchanged between MATLAB sensor and controller scripts, as well as for time-synchronization of the simulation. An interface has been developed for the scripts to be able to interact with the OMNeT++ simulation. The OMNeT++ part of the simulation is in charge of timing requirements, i.e. when to simulate the controller, wind turbine or fatigue estimator, and is thus in charge of handling the information access strategies. The wind turbine is modelled by the differential equation linearised around a chosen operating point:

$$X(t + 1) = A_d \cdot X(t) + B_d \cdot U(t) + E_d \cdot V(t) \tag{1}$$

where X is a three dimensional vector containing the state of the wind turbine, U is a two dimensional vector containing the control setpoint and V is a scalar of the wind, which is real wind data. How these interact is determined by the matrices A_d and B_d and the vector E_d, and we refer to Reference [8] for details on determining these matrices and vector. We let t be the discretized time.

4.2   Scenario Parameters

Some assumptions regarding the controller and its physical representation are taken in order to simplify the simulations. The controller is assumed to have a constant internal computation delay for calculating the set-points. At any given time there is a change in the wind turbine state it is assumed that the sensors are able to get a reading instantly. An example of this could be a new control set point or a change in wind speed. The wind turbine itself is assumed to act on a set point as soon as it is received, and we assume that this is done instantly, and that the change has an impact on the state of the wind turbine instantly. We also assume that the controller and the sensors have perfect clock synchronization. The wind data which have been used in the simulations have been downloaded from Reference [14]. The wind data used have a measurement every 12.5 ms, leading to 12 samples per control period.

The wind data set contains data covering 4000 control cycles. In order to provide confidence intervals for the results, we have split the wind data up into 10 equal intervals, each having enough wind data for 400 control cycles. It is possible to increase the number of repetitions in order to reduce the size of the confidence interval, however, this would reduce the length of the individuals runs and hence would more strongly pronounce the transient behavior caused by the simulation start.

The performance of the controller is described by two metrics: reference tracking and accumulated damage of the wind turbine. We focus on the second metric, the accumulated damage. The controller attempts to minimize this, however, a running wind turbine will always accumulate some damage. It would be expected that a delay in sensor information would lead to the controller making decisions regarding the minimization of the damage that are not optimal, which would lead to increased damage of the wind turbine.

While we cannot say there is a statistical difference between the accumulated damage of the 3G results, Fig. 8, and the WLAN results, Fig. 9, we do see a trend. It appears that the 3G results have a slight increase in accumulated damage for offsets smaller than 25 ms as we go towards 0 ms offset, while the WLAN results appear more flat. This is caused by the higher delays experienced using 3G compared to WLAN. The WLAN measurements have a mean delay (5.4 ms) that is lower than the starting offset (12.5 ms), whereas the 3G measurements have a mean delay (16.7 ms) that is higher. We also see, statistically, that the offset may not have a large impact, as long as it is chosen below a certain threshold, which in Figs. 8 and 9 is around 90 ms. The sharp increase in accumulated damage we

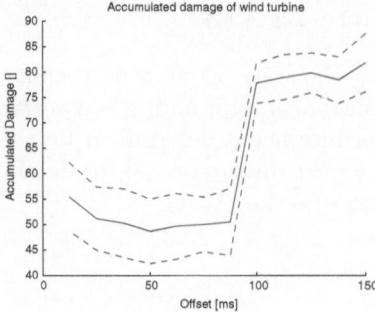

Fig. 8. Accumulated damage of wind turbine based on 3G measurements

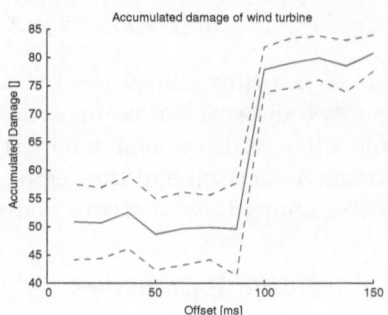

Fig. 9. Accumulated damage of wind turbine based on WLAN measurements

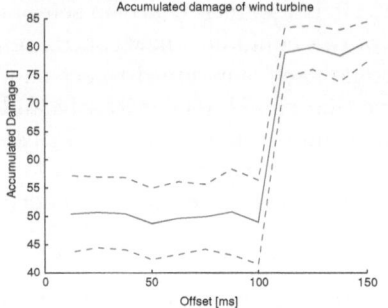

Fig. 10. Accumulated damage of wind turbine based on ideal network

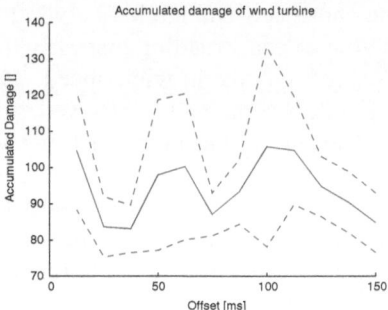

Fig. 11. Accumulated damage of wind turbine based on 2G measurements

see at this point is because this is when the wind turbine receives the control set-point. At this point the sensor information changes drastically, thus any information sent before this time will be less accurate in terms of accuracy. This offset threshold is influenced by the controller computation delay (50 ms), and the downstream delay.

For a comparison case we have also simulated a scenario where we have an ideal network. This means 0 % packet loss and 0 ms upstream and downstream delays. These results can be seen in Fig. 10, and shows that there is less variation in the accumulated damage. We further see that the offset at which the accumulated damage increases significantly has been increased to 100 ms, instead of the 87.5 for 3G and WLAN. This is because the downstream delay is 0 ms for the ideal network, thus it is only the computation delay of the controller that has an impact on this offset (Fig. 11).

The simulation results using measurements from 2G show that there is no choice in offset which is statistically better than another. We believe this to be caused by the large delays we experience on the 2G network, as the average delay is still higher than a single control period.

5   Summary

In this paper we investigated performance of a wind turbine controller when operating under different communication network technologies. The performance behavior of the different technologies was obtained from laboratory measurements of 2G, 3G, WLAN and PLC communication. We found that the 3G and WLAN communication networks provided adequate support for the controller, with WLAN showing a trend of having less fluctuations in performance. However, 2G and narrow-band PLC measurements both had one-way delays which exceeded what the controller could compensate for. This lead to the performance of the controller drastically decreasing for the simulation using 2G measurements, and it was found that using the PLC technology measurements was not possible as they lead to instability of the controller.

The analysis of wind turbine control over WLAN and 3G in Figs. 8 and 9 shows that the optimal choice of the offset parameter, i.e. the optimal scheduling of update messages by sensors, is in the interval [12.5 ms, 87.5 ms]. However, for 3G in Fig. 8, the lower bound of this interval may move up to around 20–50 ms, but the statistical fluctuations do not allow to say that with certainty. The results in this paper used extensive co-simulations to show this. An alternative way for optimized scheduling of the sensor update messages was introduced in Reference [15]. That approach uses a data quality metric, called mismatch probability, that is the probability that the controller's view on the sensor data deviates more than a certain threshold from the true sensor value at the time instant of computation. Reference [15] thereby assumed a communication network characterized by exponential delays. The investigation on how the delay traces from the actual communication technologies influence the mismatch probability models and analysis is a logical next step.

Acknowledgement. The authors would like to thank José de Jesús Barradas-Berglind for lending his wind turbine control simulation and allowing us to work with it.

This work was partially supported by the Danish Council for Strategic Research (contract no. 11-116843) within in the 'Programme Sustainable Energy and Environment', under the "EDGE" (Efficient Distribution of Green Energy) research project, and by the FP7 project SmartC2Net. FTW has been supported by the Austrian Government and the City of Vienna within the competence center program COMET.

References

1. Madsen, J., le Fevre Kristensen, T., Olsen, R., Schwefel, H.P., Totu, L.: Utilizing network QoS for dependability of adaptive smart grid control. In: ENERGYCON 2014, pp. 859–866, May 2014
2. Heemels, W., Teel, A.R., van de Wouw, N., Nesic, D.: Networked control systems with communication constraints: tradeoffs between transmission intervals, delays and performance. IEEE Trans. Autom. Contr. **55**(8), 1781–1796 (2010)
3. Bhattarai, B., Levesque, M., Maier, M., Bak-Jensen, B., Radhakrishna Pillai, J.: Optimizing electric vehicle coordination over a heterogeneous mesh network in a scaled-down smart grid testbed. IEEE Trans. Smart Grid **6**(2), 784–794 (2015)

4. Qiu, R., Hu, Z., Chen, Z., Guo, N., Ranganathan, R., Hou, S., Zheng, G.: Cognitive radio network for the smart grid: experimental system architecture, control algorithms, security, and microgrid testbed. IEEE Trans. Smart Grid **2**(4), 724–740 (2011)
5. Georgievski, I., Degeler, V., Pagani, G., Nguyen, T.A., Lazovik, A., Aiello, M.: Optimizing energy costs for offices connected to the smart grid. IEEE Trans. Smart Grid **3**(4), 2273–2285 (2012)
6. Siemens: Siemens ag (2015). http://www.industry.siemens.com/verticals/global/en/wind-turbine/communication/pages/default.aspx
7. Contemporary Control Systems, Inc.: Managed ethernet switches ensure reliable communications for a wind farm in inner mongolia, December 2010. http://www.ccontrols.com/enews/1210story2.htm
8. Barradas-Berglind, J., Wisniewski, R., Soltani, M.: Fatigue damage estimation and data-based control for wind turbines. IET Contr. Theory Appl. **9**(7), 1042–1050 (2015)
9. Ahmed, M.A., Kim, Y.C.: Communication network architectures for smart-wind power farms. Energies **7**(6), 3900 (2014)
10. 4GLTEmall: Huawei e392 4g lte fdd tdd multi-mode data card, June 2015. http://www.4gltemall.com/huawei-e392-4g-lte-multi-mode-data-card.html
11. Devolo: Devolo g3-plc modem 500k, June 2015. http://www.devolo.com/at/SmartGrid/Produkte/devolo-G3-PLC-Modem-500k
12. PC Engine: alix3d2, June 2015. http://www.pcengines.ch/alix3d2.htm
13. Mini-Box: Wistron cm9-gp minipci card, June 2015. http://www.mini-box.com/Wistron-CM9-GP-Atheros-miniPCI
14. Hansen, K.S.: Database on wind characteristics, March 2015. http://www.winddata.com
15. Madsen, J., Findrik, M., Madsen, T., Olsen, R., Schwefel, H.P.: Scheduling data collection for remote control of wind turbines. In: Energy Conference (EnergyCon) (2015). Submitted to EnergyCon 2016

Secure Communications for Ancillary Services

Matthias Krebs(✉), Stefan Röthlisberger, and Peter Gysel

School of Engineering, University of Applied Sciences
Northwestern Switzerland, Windisch, Switzerland
{matthias.krebs,stefan.roethlisberger,peter.gysel}@fhnw.ch

Abstract. Networking of distributed energy resources for ancillary services like control pooling introduces new security challenges. For economic reasons public IP networks are often used for the transport, resulting in sophisticated security requirements. Legacy devices as well as compliance with corporate network security policies must be taken into account. In this paper, we compare different communication technologies and discuss the problems of integrating legacy devices. We describe an approach that uses standardized technologies to provide secure communications for ancillary services, while at the same time requiring minimal configuration by administrators of corporate networks.

Keywords: Ancillary services · Security · Communications · Standards

1 Introduction

As far as communication in ancillary services is concerned, we consider a scenario as depicted in Fig. 1. Networked control devices, which can be controllers for different kinds of power-consuming or power-generating equipment such as heat pumps, furnaces or hydropower stations, are distributed in homes and industrial facilities. They are owned by private individuals or businesses, not grid operators. In order to participate in ancillary services, they communicate with service providers who offer services like optimization of energy consumption or control pooling. Communication takes place over corporate IP networks or even Internet, because participants can be distributed over a large area, and since the service providers are not grid operators, they have no access to communication over power lines. An other reason is that using IP is relatively inexpensive, considering most homes and businesses already have internet access.

With devices distributed across large areas and communicating over a network security becomes relevant. In order to ensure correct operation and billing, especially when participating devices are remotely controlled, they must be authenticated and data sent over the network must be protected against manipulation. Furthermore, critical systems like the provision of electrical power should not be left vulnerable to cyber attacks.

While new equipment designed for networking, such as energy management devices and load controllers, is generally being installed in new buildings or after

© Springer International Publishing Switzerland 2015
S. Gottwalt et al. (Eds.): EI 2015, LNCS 9424, pp. 141–152, 2015.
DOI: 10.1007/978-3-319-25876-8_12

fundamental renovation, many companies and homes already possess existing control devices. In some cases these devices can already be considered networked in some way, because they feature interfaces accessible via IP like Modbus [20], thus providing some level access to other systems. It is not to be expected that functional equipment is thrown away and replaced without necessity, after all considerable investments may have been made and should be amortized before new equipment is purchased.

Therefore, the integration of existing devices has to be taken into account, despite the fact that these devices might not fulfill the security standards necessary for communication over Internet. In that case, the devices are considered legacy. Integrating these legacy devices introduces additional challenges, because not only does it require secure communication with the outside world, but also protection of the local network.

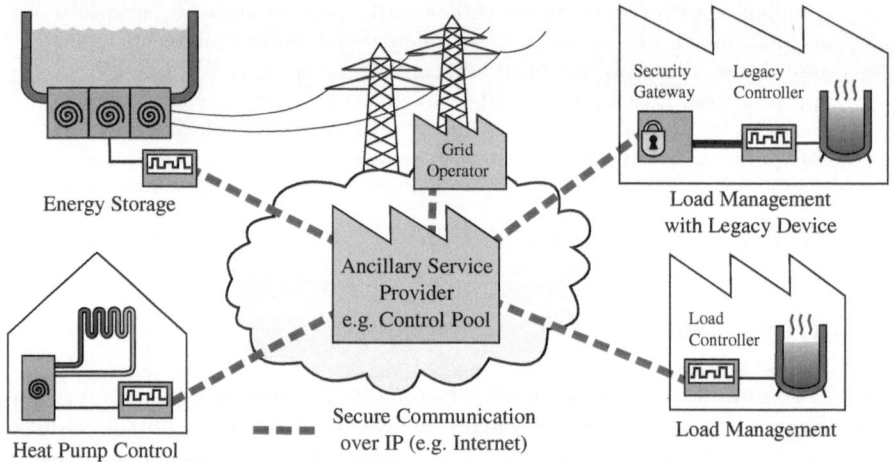

Fig. 1. Ancillary Services in a wide-area scenario

This paper focuses on the challenges of finding a secure IP communication approach that is suitable for ancillary services and complies with security requirements of corporate networks. In Sect. 2, we discuss related work. The challenges which arise when deploying a device in a corporate network are discussed in Sect. 3. We then compare potential protocols in Sect. 4. A use case of a distributed load management system is presented in Sect. 5. We conclude our paper with a reflection about the feasibility of the chosen approach and future developments.

2   Related Work

The German Federal Office of Information Security has worked out a protection profile for smart meter gateways [1]. The document describes security objectives and their requirements. Such a gateway has a connection to the Internet and accesses smart metering devices locally. The gateway's job is to collect, process and transfer data from the attached meters. In terms of security, the goals are protecting the privacy of the consumers, ensuring a reliable billing process and protecting the power grid as a whole. An important requirement is the usage of a security module (e.g. a smart card) [2] providing various functions related to encryption and authentication. A second premise is that all devices communicating with the gateway have to use encryption and mutual authentication, thus making the integration of legacy devices impossible.

The Internet Engineering Task Force (IETF) has published a document on how Internet communication protocols could be used with networked energy resources. It has been released as RFC 6272 [13]. The document covers virtually all aspects of networking, including network topology, secure communication and different application protocols. Since it can be thought merely of a reference, not an actual communication standard, the document does not provide concrete recommendations on which communication approaches should be chosen.

The design of a secure access gateway for home area networks is considered in [3]. Their article focuses on secure, real-time remote monitoring and control of managed devices using a smart phone. The proposed system architecture enables the managed devices to send alerts to the smart phone. The emphasis is on physical layer security of wireless networks (e.g. OFDM and GSM) and capacity challenges therein.

The European Telecommunications Standard Institute (ETSI) has approved a communication standard [8] for networked energy resources, the Open Smart Grid Protocol (OSGP) [7]. The standard specifies networking protocols and data models for the data transfer of smart meters. The default approach is communication over a power line channel (PLC), although it does not depend on a specific physical layer. Custom cryptography methods are used for security.

The authors of [4,5] consider using the WebSocket protocol [14] in machine-to-machine communications. The former focuses on electric vehicles communicating over cellular networks, whereas the latter describes a gateway that accesses a wireless sensor network and forwards the collected data through WebSocket.

3   Challenges

Searching for an approach to secure communication for ancillary services, we have identified a number of challenges. We explain these challenges in the following paragraphs.

Secure Communication becomes important whenever it takes place over an insecure channel. Current smart meters, as an example, often communicate through the power line (PLC) and not IP because the electricity provider can access it. Since everyone could tap into a power line PLC can be considered insecure. The same applies to devices that communicate over an IP network, because the data packets are routed across a geographically distributed infrastructure and could be tampered with anywhere along their way. It is therefore essential that all the data exchanged between a networked device and a service provider is encrypted end-to-end, so that no manipulation can occur. It is also important that the service provider can be certain it is talking to the proper device, thus an authentication and identity assertion method is necessary. There must be no direct remote access to the participating devices, and the service provider's infrastructure must be secured in particular, because a potential attacker could gain access to all connected devices. We do not tackle denial-of-service attacks as they can never be fully prevented.

Compliance with corporate network security policies is important when a device is to be integrated into a corporate network. Many companies employ restrictive policies regarding network security. This means that they are not willing to lessen their security policy just for one device. Such companies generally use at least a firewall which blocks all inbound traffic from the Internet by default. Even more restrictive configurations include blocking most TCP/UDP ports from the corporate network to the Internet, except for very common ports (HTTP, HTTPS). A networked device has to respect that and provide means to operate under these circumstances. Inbound connections should thus be avoided whenever possible. Additional obstacles are introduced by the utilization of proxy servers that cache data, restrict and filter access to the Internet. These often limit access to web protocols like HTTP and HTTPS and do not allow other protocols to pass. A threat to secure communication is deep packet inspection (DPI), which some firewalls or proxy servers perform. It inspects even encrypted traffic by decrypting, analyzing and re-encrypting the packets prior to forwarding. A custom certification authority (CA) is declared trusted on the clients inside the corporate network, and instead of presenting the real server certificate to clients when they open an HTTPS web site, a certificate signed by the custom CA with the same server name is given. Because the clients trust the custom CA, they accept the forged certificate. The result is encrypted communication between the client and the proxy server, but the latter is able decrypt the data. After analysis and approval of the client's messages, the proxy then forwards them to the actual server, now using the real server certificate for encryption. The actual problem is that this qualifies as a man-in-the-middle attack, as the proxy could manipulate the data. End-to-end encryption would be impossible in such a situation.

Uncomplicated configuration is desirable, because the installation of a new device should not force the administrator to configure complex firewall rules or even open inbound ports. Using standardized Internet protocols on standard ports can circumvent this problem.

The *integration of legacy devices* adds an extra layer of complexity, because by our definition a legacy device itself cannot communicate in a secure fashion. Therefore, the insecure communication must be encapsulated by a secure communication protocol, which is then used for data exchange over the Internet. This problem can be addressed through the introduction of additional infrastructure, such as a virtual private network (VPN) gateway that routes any network traffic over an encrypted tunnel [6], or a security gateway (SGW) as depicted in Fig. 2, which exclusively has access to the Internet and acts as a protocol converter encapsulating the legacy device's messages.

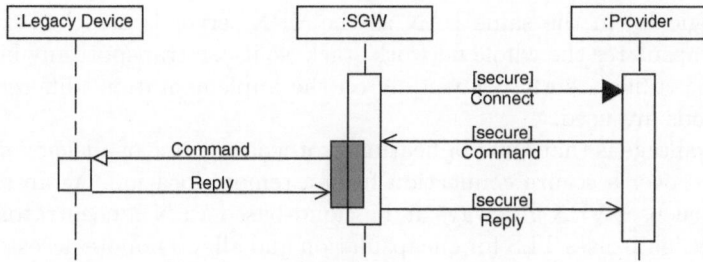

Fig. 2. Communication between a service provider and a legacy device, using a security gateway as intermediary

4   Evaluation of Communication Protocols

To find a suitable approach, we have based our evaluation on the following five requirements for communication between a device and the service provider. They have been chosen to allow for seamless adoption into a corporate environment.

1. Efficiency, because bandwidth may be limited (e.g. GSM/GPRS)
2. End-to-end encryption and integrity protection with mutual authentication
3. No obligation for firewall modifications, especially not opening a port to allow Internet traffic into the company's network
4. Ability to pass through intermediary proxy servers
5. Bi-directional communication ability.

4.1   Security

We have considered two approaches to secure the communication channel, independent of the transport protocol to be used.

Transport Layer Security. The Transport Layer Security (TLS) protocol [11] provides privacy, data integrity and authentication using a combination of asymmetric and symmetric cryptography. When a connection is established, a handshake is performed first, using certificates containing a public and private key to exchange a symmetric key that will be used to encrypt the data packets (e.g. AES-256). By obfuscation of the transmitted data through encryption, privacy is guaranteed. Message authentication codes (MAC) prevent message tampering and forgery. At least a server certificate is mandatory for an encrypted connection, and an optional client certificate can be given to act as authentication credentials. TLS adds a transparent security layer to any TCP connection, thus being compatible with any transport protocol built upon TCP.

Virtual Private Network. A VPN offers a solution to interconnect two remote networks through a bi-directional encrypted tunnel. A client connected to a

VPN is logically in the same LAN as the VPN server is attached to. A VPN tunnel encapsulates the whole network stack, so it can transport any kind of network traffic transparently. Depending on the implementation, different encryption methods are used.

An advantage is that any application protocol, e.g. one of a legacy device, can be accessed over a secure connection from a remote location. As an example, a solution that uses VPN gateways and a cloud-based VPN concentrator is offered by ADSTec [6]. It uses TLS for encapsulation and allows remote access to devices in a local network.

There are some disadvantages, too. The encapsulation of the whole network stack results in considerable overhead, which adds to the overhead introduced by TLS. Without careful configuration, new security holes may be introduced as well, because unlike a TCP connection using TLS, whose encapsulated channel remains a single point-to-point connection, a VPN can link entire network segments and thus has the potential to expose services that should not be exposed.

4.2 Communication Protocols

Our goal is to find an efficient approach which can be used without compromising the security of a corporate network. We have analyzed several communication protocols. In the following sections, we compare them in terms of efficiency and how well they meet the previously mentioned challenges. An overview of the discussed protocols and their fulfillment of the requirements can be found in Table 1.

HTTP Polling and Long Polling. The Hypertext Transfer Protocol (HTTP) is a widely used request-response-based protocol used to communicate with web servers. We have looked at two variants. *Polling* (Fig. 3a) repeatedly sends requests to the server. The server immediately responds either with new information or an empty response. The second variant is *Long Polling* (Fig. 3b). The main difference is that Long Polling does not send empty responses back to the client, instead the connection is kept open until a request can be answered with new information.

Server-Sent Events. *Server-Sent Events* are currently being standardized as part of HTML5 by the W3C [12]. It offers a light-weight approach to push messages from the server to the client. The client initiates the connection which is basically an HTTP GET request with the *Content-Type* header set to *text/event-stream*. The server keeps the connection open and sends (pushes) multiple messages to the client until the connection is explicitly closed by the server or the client.

WebSocket. *WebSocket* (see Fig. 3c) is a bi-directional protocol using a single socket for communication. It is specified in RFC6455 [14] and is a W3C working draft [15]. A WebSocket client establishes a connection using the HTTP *upgrade* header during the initial handshake. The HTTP connection is then upgraded to a

WebSocket connection. After that, it is no longer considered an HTTP connection. After being established, the connection persists until a participant closes it explicitly. WebSocket provides message-based communication with minimal overhead.

Raw TCP Sockets. An established TCP connection (see Fig. 3c) provides a bi-directional communication channel that transports a stream of binary data. Tasks like distinguishing individual messages have to be adopted by a higher-level application protocol, as a raw TCP connection provides no such means itself. Examples of protocols built upon TCP sockets are XMPP or SIP, as specified in RFC6272 [13].

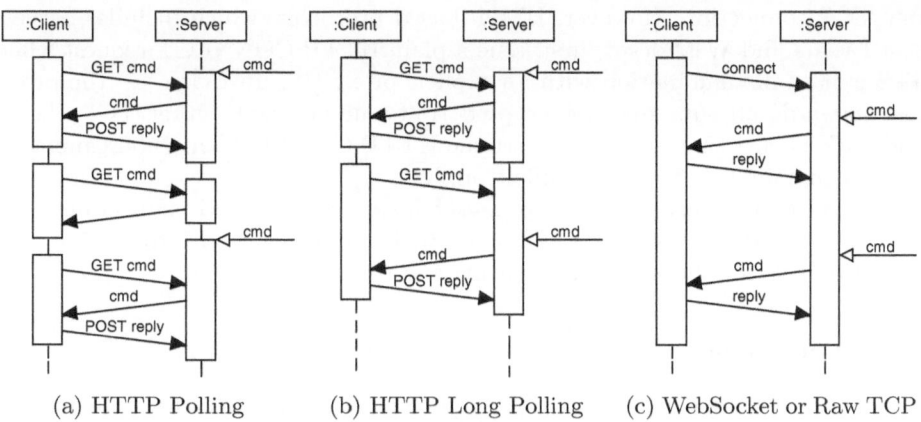

(a) HTTP Polling   (b) HTTP Long Polling   (c) WebSocket or Raw TCP

Fig. 3. Communication between client and server

4.3 Firewall Friendliness

Assuming that a firewall allows outgoing TCP connections on all ports from the company's network to the Internet, all of the described protocols will work without a hassle. In contrast, if the firewall is restricted to only allow Internet access over standard HTTP ports (80 for plain HTTP, 443 for HTTPS), which is likely in many corporate networks, an application protocol based on a raw TCP socket cannot be used if it is configured to use a different port.

Transport protocols based on HTTP are usually configured to use a standard port by default, so no modifications have to be made to the firewall. If a protocol uses these ports but is in fact not based on HTTP, it might be blocked by a packet-inspecting firewall. Protocols like WebSocket circumvent this problem by using an actual HTTP request to initiate the connection and then perform an HTTP Upgrade to switch to the actual protocol, so the firewall treats it as HTTP-based.

VPN implementations using TLS, like OpenVPN, can operate on a single socket on port 443. Other technologies, like IPsec, may require a dedicated port to be opened on a firewall and sometimes enabling specific configuration options such as *VPN pass-through*, or even inbound ports. This would contradict requirement 3.

4.4   Proxy Server Traversal

Compatibility with proxy servers depends on the kind of proxy being used. A *transparent* proxy is integrated into the network so that traffic is automatically routed through the proxy, with a client being unaware of its existence. The proxy usually just forwards traffic and might provide caching or filtering in the case of plain HTTP.

An *explicit* proxy is a different story. A client must be configured to use the proxy, or it does not get any Internet access. The proxy is limited to HTTP connections, so in theory all HTTP-based protocols should work and raw TCP connections should fail. TLS should fail, too, because to the proxy it looks like a raw TCP connection. However, HTTP-based protocols, which includes Server-sent Events and WebSocket, first issue a plain HTTP CONNECT request when TLS is used in combination with an explicit proxy [18]. In that case, the proxy just forwards all subsequent encrypted traffic unmodified. Other TCP clients and VPN clients that do not support the HTTP CONNECT method cannot be used in combination with an explicit proxy.

Proxy servers performing deep packet inspection on TLS connections prevent end-to-end encryption (requirement 2). The only solution is to configure an exception rule for those clients requiring end-to-end encryption.

4.5   Performance

Performance is influenced by several factors. One is protocol overhead. Regarding HTTP, request and response headers are a common cause of overhead. The amount of header information heavily depends on the application, it can take from 100 bytes to more than 1 KB for each request or response. We assume a scenario where commands may be issued from the server to the device, so when using HTTP Polling, the device has to poll the server with HTTP GET requests to check for new messages. If there is a message it is included in the HTTP response body. Otherwise, the response contains just overhead. Messages from the device are sent to the server as HTTP POST requests.

HTTP Long Polling is similar, but the device polls and then waits for a response, so there is less overhead generated than through continuous polling.

When using Server-Sent Events, the server can push messages to the device, because a persistent uni-directional connection from the server to the device is established. Nevertheless, the device has to use additional HTTP POST messages to send data back to the server, since Server-Sent Events do not offer a bi-directional channel.

WebSocket requires just one HTTP request to initiate the connection, after that it is a bi-directional and persistent communication channel, similar to a raw TCP socket. Each WebSocket message has an overhead of 2 to 12 bytes, depending on the payload size.

In [19] the authors compare HTTP Polling with WebSocket in terms of protocol overhead. The HTTP header they use is 871 bytes long, and a WebSocket

Table 1. Comparison of protocol capabilities

	Efficient	TLS	No firewall setup needed	Proxy server pass-through	Bi-directional
HTTP Polling	X	✓	✓	✓	X
HTTP Long Polling	X	✓	✓	✓	X
Server-sent Events	X	✓	✓	(✓)	X
WebSocket	✓	✓	✓	(✓)	✓
Raw TCP socket	✓	✓	(✓)	X	✓

message has an overhead of 2 bytes. They test 1'000, 10'000 and 100'000 simultaneous requests per second to illustrate the difference. At 10'000 requests per second HTTP Polling uses 66 Mbps, compared to 0.153 Mbps using WebSocket. In a comparison between HTTP Long Polling, WebSockets and raw TCP sockets, the TCP socket was always faster and had more throughput than the others [17]. The larger the payload was, the larger the difference became. Similar results are shown in [4], where the bitrate of WebSocket is measured 60–70% lower than HTTP Polling, and TCP being the most efficient.

Besides protocol overhead, the use of TLS encryption introduces additional overhead. Establishing the connection can require 6 to 10 KB, because certificates have to be exchanged during the hand-shake. The exact size depends on the cipher suites being used and the certificate's key length. In an open connection, overhead of an encrypted data packet would not exceed 60 bytes (including a maximum of 31 bytes for AES-256 padding). Thus TLS shows best efficiency in persistent connections.

In cases where a VPN connection is chosen to secure communication, further overhead is generated. A TLS-based VPN like OpenVPN has to encapsulate the full network stack, including IP and TCP or UDP headers of each packet to be encapsulated. These packets are then broken down into TLS records and again into TCP packets.

An other performance indicator of a communication channel is latency. In [16] the authors compare the latency of HTTP Polling, HTTP Long Polling and WebSocket. With polling it is measured 2.3 to 4.5 times higher than with WebSocket. HTTP Long Polling has achieved both lower and higher latency in comparison to WebSocket, depending on the situation. Over the longest distance (Canada to Japan), the average latency of WebSocket is 3.8 to 4.0 times lower compared to HTTP Long Polling.

4.6 Decision

After comparing the capabilities of the protocols we analyzed, we have decided to go with WebSocket, as it fulfills our requirements in Table 1 best. It offers efficient bi-directional communication and plays well even in corporate environments. Its high level API provides connection control and message handling, thus an application protocol does not need to implement these functions. Security can be implemented using standardized TLS encryption and authentication without much additional overhead, because WebSocket is already designed for persistent connections, which is also the most efficient scenario for TLS.

We have decided against VPN for multiple reasons. Since VPN is designed to interconnect whole networks, malicious traffic could be injected. This happened in 2003 [10], where the SQL slammer worm propagated through VPNs. A similar situation could happen if the service provider's endpoint is compromised, whereby an attacker would potentially gain access to all client networks connected to the endpoint. Complex configuration is necessary, for example with IPsec, which requires specific firewall settings and inbound connections, Also, a VPN approach adds more complexity in case of a single device, which most likely just uses a single application protocol. When legacy devices are to be integrated, we opt for a security gateway (SGW) that provides a secure connection and converts the legacy device protocol.

We have also looked at existing standards such as OSGP. One problem is that it is designed for PLC, while we need an approach that works over IP networks. The major problem, however, is security. A cryptographic analysis [9] has found several weaknesses, such as non-standard digest functions and the use of RC4. Due to the security issues we would not use OSGP in its current form.

5   Use Case of a Distributed Load Management System with Legacy Devices

In a project funded by the Swiss Federal Commission for Technology and Innovation (CTI) we have developed a prototype of a distributed load management system for control pooling. The load controllers are of legacy kind. They are already widely deployed and still perfectly capable of performing their duty, but due to technical limitations they cannot be upgraded with a secure communication method. Hence they are not suitable for Internet communication.

To address this problem, we have developed a security gateway (SGW), which is an embedded device that is installed in the customer's corporate network together with the load management device. The SGW subsequently provides a secure communication channel to a centralized control service acting as the service provider. The actual data exchange with the load controllers remains in the customer's network. Based on our evaluation we have chosen WebSocket for communication between the SGW and the control service. The SGW device features an ARMv7 processor and is based on an embedded Linux platform. The gateway software is implemented in Java running on Java SE Embedded. The control service is also implemented in Java and thus shares part of its code base with the SGW. Messages exchanged between the SGW and the control service are encapsulated using a custom data model based on compact XML. We have chosen this approach for reasons of flexibility.

The secure connection employs TLS v1.2 with AES-256 [11]. We use the mutual authentication feature of TLS, giving each SGW its own certificate. The certificates are signed by a custom certificate authority (CA) created solely for use in our system. Since our CA is the only one trusted by the SGW and control service, we can ensure that man-in-the-middle attacks are not possible. Certificates signed by a different CA would automatically be rejected. All certificates can be revoked through a certificate revocation list (CRL), which is queried by the SGW and the control service upon establishing a connection. This allows us to deny access in case a device or its certificate is stolen.

We have conducted feasibility tests in a real-world environment, where communication effectively takes places over a public Internet connection. The load controllers were accessed through the SGW five times a second, with an XML message of 480 bytes. Results have shown that the SGW's processor can easily handle the TLS-encrypted WebSocket protocol. Round-trip time is mostly affected by the number of hops and the connection with the lowest bandwidth. Sending 480 bytes over a 20 kbps GPRS connection would take 192 ms and 9.6 ms over 400 kbps DSL. We achieved a total round-trip time of 50 ms between the control service and controller using cable or fiber connections. We have successfully tested the system with an UMTS connection as well. Even with a GPRS connection, those five message could still be sent in less than a second. The control service performance was tested with software clients and was able to handle at least 1000 concurrent connections.

6   Conclusions and Future Work

During the development of our prototype, we have decided to go with the approach of using the WebSocket protocol in combination with TLS encryption and mutual authentication. We have considered it a feasible approach, because it is based on already standardized technologies and can be easily implemented, works with all kinds of IP connections and performs well in terms of bandwidth and latency. Network configuration is simple, as it does not have higher requirements than Internet access for web browsers. This, and the guaranteed end-to-end encryption will help gain the acceptance of potential customers. To integrate legacy devices, we have found that an SGW is a good solution. It is inexpensive, easy to install and requires little configuration.

Currently, we use our prototype system in a flagship project, where we use an SGW to remotely access the process control system of some water suppliers in Switzerland. Their pump strategy is optimized on a regular basis to reduce energy costs and provide tertiary control energy, while still maintaining a reliable water supply.

Development of our solution is still in progress. Future research includes further improvement of security and reliability, secure management of certificates and secure remote administration and maintenance of devices.

References

1. Federal Office for Information Security, Germany. Protection Profile for the Gateway of a Smart Metering System (Smart Meter Gateway PP). Version 1.2, 18 March 2013. https://www.bsi.bund.de/DE/Themen/SmartMeter/Schutzprofil_Gateway/schutzprofil_smart_meter_gateway_node.html
2. Federal Office for Information Security, Germany. Protection Profile for the Security Module of a Smart Meter Gateway (Security Module PP). Version 1.0, 18 March 2013.https://www.bsi.bund.de/DE/Themen/SmartMeter/Schutzprofil_Security/security_module_node.html
3. Li, T., Ren, J., Tang, X.: Secure wireless monitoring and control systems for smart grid and smart home. IEEE Wireless Commun. **19**(3), 66–73 (2012)
4. Pérez, J., Nurminen, J.K.: Electric vehicles communicating with WebSockets - measurements and estimations. In: 4th IEEE PES Innovative Smart Grid Technologies Europe (ISGT Europe), Copenhagen, 6–9 October 2013
5. Shuang, K., Shan, X., Sheng, Z., Zhu, C.: An efficient ZigBee-WebSocket based M2M environmental monitoring system. In: 2014 IEEE 12th International Conference on Dependable, Autonomic and Secure Computing (2014)
6. ADS-Tec Big-Linx Remote Service Cloud using OpenVPN. http://www.ads-tec.de/industrial-it/cloud-big-linx/big-linx.html
7. The Open Smart Grid Protocol Alliance. http://www.osgp.org/
8. ETSI Approves Open Smart Grid Protocol (OSGP) for Grid Technologies. http://www.etsi.org/news-events/news/382-news-release-18-january-2012
9. Jovanovic, P., Neves, S.: Dumb Crypto: Practical Cryptanalysis of the Open Smart Grid Protocol. Cryptology ePrint Archive, Report 2015/428 (2015)
10. North American Electric Reliability Council: SQL slammer worm lessons learned for consideration by the electricity sector, 20 June 2003. http://www.utexas.edu/law/journals/tlr/sources/Issue90.1/Thompson/NAERCSlammerReport.pdf
11. Dierks, T., Rescorla, E.: The Transport Layer Security (TLS) Protocol Version 1.2., August 2008. http://tools.ietf.org/html/rfc5246
12. Hickson, I.: Server-Sent Events, W3C Candidate Recommendation, 11 December 2012. http://www.w3.org/TR/eventsource/
13. Baker, F., Meyer, D.: Internet Protocols for the Smart Grid, June 2011. http://tools.ietf.org/html/rfc6272
14. Fette, I., Melnikov, A.: The WebSocket Protocol, December 2011. http://tools.ietf.org/html/rfc6455
15. Hickson, I.: The WebSocket API, W3C Candidate Recommendation, 20 September 2012. http://www.w3.org/TR/websockets/
16. Pimentel, V., Nickerson, B.G.: Communicating and displaying real-time data with WebSocket. IEEE Internet Comput. **16**, 45–53 (2012)
17. Agarwal, S.: Real-time web application roadblock: performance penalty of HTML sockets. In: IEEE ICC 2012 - Communication QoS, Reliability and Modeling Symposium, pp. 1225–1229 (2012)
18. Lubbers, P.: How HTML5 Web Sockets Interact With Proxy Servers, 16 March 2010. http://www.infoq.com/articles/Web-Sockets-Proxy-Servers
19. Lubbers, P., Greco, F.: HTML5 Web Sockets: A Quantum Leap in Scalability for the Web, March 2010. http://soa.sys-con.com/node/1315473
20. The Modbus organization: The Modbus standard specification. http://www.modbus.org

Cyber Security Analysis of Smart Grid Communications with a Network Simulator

Roberta Terruggia[✉] and Giovanna Dondossola

Transmission & Distribution Technologies Department Ricerca
Sul Sistema Energetico, Milan, Italy
{Roberta.Terruggia,Giovanna.Dondossola}@rse-web.it

Abstract. This paper proposes a methodology for analyzing communication security in smart grid domains. As an application case, in this paper, we focus on information exchanges between Distributed Energy Resources (DER) and primary substations employed in the medium voltage control of active grids. The ICT (Information and Communication Technology) architecture of the application is modeled through a network simulator integrating the standard communication modules and the attack processes experimentally evaluated in a voltage control cyber security testbed. The paper presents the cross validation of the simulation model with the experimental traces, the sensitivity analysis with the validated model of the flooding attack effects on the communication performance and the scaling-up capability of the simulation model for the analysis of more realistic grid size.

Keywords: Cyber security · Smart grid · DER · Communication · MMS

1 Introduction

The evolution of the energy infrastructures connecting Distributed Energy Resources (DER) at different levels of the power grids leads to rethink the functionalities of the Supervisory Control And Data Acquisition (SCADA) systems. The role of the ICT (Information and Communication Technology) infrastructure is becoming more and more fundamental and the communications among power control entities become crucial assets of the grid operation. Such ICT architectures of future smart grids may be, especially in the case of DER, based on heterogeneous and third party telecommunication services and so prone to cyber attacks. This advanced scenario pushes the cyber security issues and the need of managing ICT risks at a top position in the smart grid agenda. The innovative idea driving the research activity presented in the paper is to use an ICT network simulator for developing a sufficiently accurate communication model providing reliable evaluations of the cyber attack effects on the legal communications. The accuracy of the simulation models is granted by the usage of implementation and data from a laboratory testbed. In this paper, we describe the integration of

© Springer International Publishing Switzerland 2015
S. Gottwalt et al. (Eds.): EI 2015, LNCS 9424, pp. 153–164, 2015.
DOI: 10.1007/978-3-319-25876-8_13

a real traffic protocol and an application layer taken from a voltage control test-bed into a simulation model in order to obtain an aligned behaviour. The model can be validated by comparing the outcome from the simulation runs with those from the real setup reinforcing the degree of confidence in the selected approach. Once the model is validated through the testbed comparison, we are allowed to scale up the model and include in the simulation more complex scenarios in order to perform larger scale analysis not addressable within the testbed size limits. The paper is structured as follows: Sect. 2 introduces an overview on network simulators used in smart grid domains. Section 3 describes the Medium Voltage Control function, its communications and protocols from the testbed. The model of communication addressed in this paper is presented in Sect. 4, where the simulation models integrating malicious traffic from the testbed are described. In Sect. 5 some preliminary results from the attack analysis are discussed. Section 6 concludes the paper and presents the future work.

2   Related Work and Tools

Different simulation tools exist that can be used to model the ICT architecture of a power grid. In the following we provide an overview of the mostly used. OPNET [1] is a commercial tool and the main applications of this tool are planning, optimization and evaluation of large (corporate) networks that need to design, analyze or rearrange communication network systems. OPNET in smart grid domain is used in particular for hybrid simulation of power systems and ICT for real-time applications [2] and for co-simulation [3]. Another commercial network simulator is QualNet [4]. In the context of smart grid applications several approaches using QualNet are discussed in the literature [5]. The simulation focus is the analysis of protocols on a large scale (nodes in the tens of thousands) in heterogeneous (unicast, multicast, satellite, internet) networks. A drawback of QualNet emanates from the fact that it is focused on the analysis of protocols and their interaction with each other and the network structure, which makes it difficult to quantify the impact of ones own application on the network. Another network simulation tool is OMNeT++ [6]. It is a modular, extensible and component-based open source simulation library and framework, primarily employed for building network simulators. This framework has been used for modeling communication networks and distributed systems for smart grid applications in [7]. Other well-known open-source network simulators are the Network Simulator 2 (NS-2) [8], and its successor, Network Simulator 3 (ns-3) [9]. ns-2 and ns-3 are object-oriented, discrete event-controlled communication network simulators which are used for research and development. The development of the NS-2 has been abandoned in 2006 in favour of a new product development, i.e. ns-3. ns-3 is a free, open-source software, licensed under GNU GPLv2, written in C++ and offering Python bindings. It is possible to describe the architecture of the network and to schedule simulation events that will be executed at defined time. Different protocols (TCP, UDP, RIP, OSPF, etc.), data traffic sources (FTP, Telnet, Web, CBR, VBR, etc.), mechanics for router queue management as well as of routing algorithms can be included in the simulation model. There are also implementations for the MAC-layer and multicast-protocols for wired and wireless communication networks. The simulation results can be saved to a trace-file and can be analyzed with self-written scripts or with specific thirty party tools. An important utility introduced in ns-3 is the Direct Code Execution (DCE) functionality. This provides the possibility to include and execute existing implementations of protocols and applications without the need of rewriting the source code. This has been the reason why we have selected it as a good candidate tool for our research purposes. NS-2/3 are used in order to evaluate ICT infrastructures for the smart grids, e.g. for the improvement of existing infrastructures based on 802.11s Mesh Networks for Smart Metering [10], or in order to evaluate the interaction between ICT and energy networks in co-simulation as discussed in [11]. Thanks to the DCE functionality of ns-3, in this paper we propose a simulation analysis approach that addresses the cyber security evaluation of the smart grid communications considering the real applications exchanging traffic conform to the communication standards used in the power grids.

3   Medium Voltage Control Testbed

The connection of a consistent amount of DERs to medium voltage grids can influence the status of the whole power grid. In particular it is possible to see effects on the capacity of the DSO (Distribution System Operator) to comply with the contracted terms with the Transmission System Operator and so on the quality of service of their neighbour grids. This difficulty not only could be transferred into charges to the DSO, but it may also impact on the TSO operation because the scheduled voltages at grid nodes could not be observed and voltage stability problems cannot be managed properly. In order to maintain stable voltages in the distribution grids a Voltage Control function has been specified in [12] whose main functionality is to monitor the grid status from field measurements and to compute optimized set points for DERs, flexible loads and power equipment deployed in HV(High Voltage)/MV(Medium Voltage) substations. The Voltage Control is a function of the SCADA node of a HV/MV substation control network [12]. In order to compute an optimized voltage profile the algorithm involves communications through components inside the DSO area, but also exchange of information with systems outside the DSO domain. In particular DERs communicate with the substation SCADA via a DER Control Network, possibly deploying heterogeneous communication technologies available in different geographical areas. DERs periodically provide measurements to the substation SCADA and receive set points in order to keep optimized voltage profiles. The RSE PCS-ResTest Lab (Power Control Systems Resilience Testing Laboratory) [17] hosts a test platform for running cyber security experiments over realistic control scenarios, implementing the message exchange involved in the Medium Voltage Control using the IEC 61850 standard with MMS (Manufacturing Message Specification) profile [14]. More info on the testbed architecture can be found in [15].

4   Simulation Model

In this section the ns-3 tool introduced in the previous section is used for building the communication architecture. In the following subsection the different developed models are described.

4.1   Communication Model

Figure 1 presents a sketch of the model architecture for the legal communications. The network is composed by 7 main nodes: at DER site, the DER server (1) and the router (2); at HV/MV substation site, the router (3) for DER communication, the SCADA server (4) and the router (5) for center communication; at center level, the router (6) and the SCADA server (7). As in the testbed, the router nodes have multiple network interfaces, one for each network link.

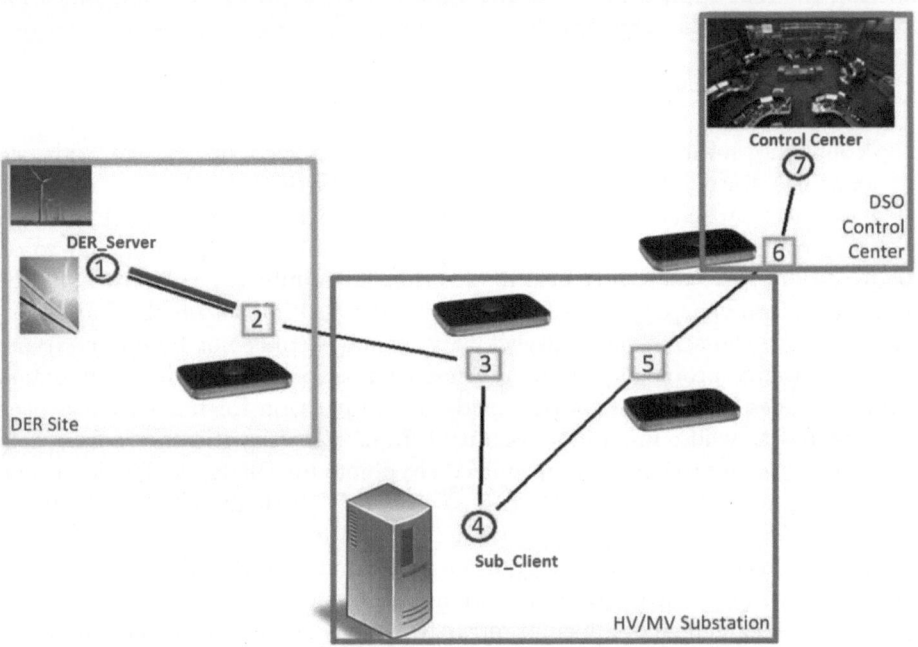

Fig. 1. Logical architecture of the simulation model

In particular we focus on DER primary substation communication and considering the Voltage Control function we have two types of information flows: periodic data representing the measurements from the DER site to the primary substation and possible asynchronous set-point sent by the primary substation in order to control the DER behavior. Both the information flows use the MMS protocol over an always on TCP/IP connection. The correspondent ns-3 model is again composed by 7 nodes belonging to 5 different networks: DER LAN NW, DER-SUB WAN NW, SUB LAN NW, SUB-CENTER WAN NW and CENTER LAN NW. Each node has an IP address depending on the network where it is placed. A sketch of the network addressed is presented in Fig. 2.

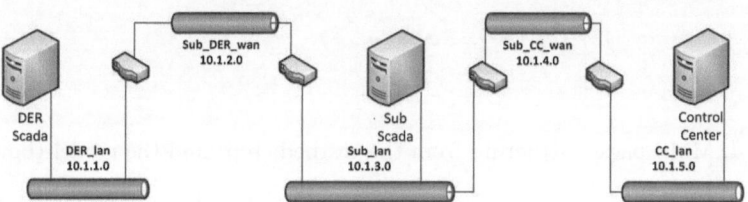

Fig. 2. Model architecture

At DER server (node 1) a MMS server application is installed: using the DCE feature of ns-3, we install in the model the testbed application based on the libiec61850 library [16]. At SCADA server (node 4) the corresponding MMS client application is running. We choose to use the testbed implementations instead of ns-3 built-in applications in order to obtain more realistic traffic: the reports containing DER measurements are sent every 2 seconds and if we sniff the traffic of the simulation we expect to obtain the same trace of the real application with the same traffic profile (packet size and content of IP, TCP and higher layer protocols in the MMS stack). In Fig. 3 the comparison between the trace of the information flow from the RSE testbed and the simulation model is shown. It is possible to see the exchange of measurements (the MMS reports) from the DER SCADA to the primary substation SCADA. The traffic pattern is the same; this means that the information flow emitted by the simulator is fully comparable with the real application information flow. Figure 4 allows comparing the structure of a MMS report packet containing the DER measurements. It is possible to note the full stack having the MMS protocol at the highest level.

```
48 14.964851    158.47.121.32 192.168.1.43  TCP  66 35314 > iso-tsap [ACK]     62 17.002374  10.1.1.1 10.1.3.1  MMS   unconfirmed-PDU
49 16.978357    192.168.1.43  158.47.121.32  MMS  179 unconfirmed-PDU           63 17.204617  10.1.3.1 10.1.1.1  TCP   49153>102 [ACK]
50 16.978454    158.47.121.32 192.168.1.43  TCP  66 35314 > iso-tsap [ACK]     64 19.002374  10.1.1.1 10.1.3.1  MMS   unconfirmed-PDU
51 18.990947    192.168.1.43  158.47.121.32  MMS  179 unconfirmed-PDU           65 19.204617  10.1.3.1 10.1.1.1  TCP   49153>102 [ACK]
52 18.991044    158.47.121.32 192.168.1.43  TCP  66 35314 > iso-tsap [ACK]     66 21.002374  10.1.1.1 10.1.3.1  MMS   unconfirmed-PDU
53 21.004561    192.168.1.43  158.47.121.32  MMS  179 unconfirmed-PDU           67 21.204617  10.1.3.1 10.1.1.1  TCP   49153>102 [ACK]
54 21.004657    158.47.121.32 192.168.1.43  TCP  66 35314 > iso-tsap [ACK]     68 23.002374  10.1.1.1 10.1.3.1  MMS   unconfirmed-PDU
56 23.007143    192.168.1.43  158.47.121.32  MMS  179 unconfirmed-PDU           69 23.204617  10.1.3.1 10.1.1.1  TCP   49153>102 [ACK]
57 23.007238    158.47.121.32 192.168.1.43  TCP  66 35314 > iso-tsap [ACK]     70 25.002374  10.1.1.1 10.1.3.1  MMS   unconfirmed-PDU
58 25.022732    192.168.1.43  158.47.121.32  MMS  179 unconfirmed-PDU           71 25.204617  10.1.3.1 10.1.1.1  TCP   49153>102 [ACK]
59 25.022828    158.47.121.32 192.168.1.43  TCP  66 35314 > iso-tsap [ACK]     72 27.002374  10.1.1.1 10.1.3.1  MMS   unconfirmed-PDU
60 27.036353    192.168.1.43  158.47.121.32  MMS  179 unconfirmed-PDU           73 27.204617  10.1.3.1 10.1.1.1  TCP   49153>102 [ACK]
61 27.036449    158.47.121.32 192.168.1.43  TCP  66 35314 > iso-tsap [ACK]     74 29.002374  10.1.1.1 10.1.3.1  MMS   unconfirmed-PDU
62 29.038935    192.168.1.43  158.47.121.32  MMS  179 unconfirmed-PDU           75 29.204617  10.1.3.1 10.1.1.1  TCP   49153>102 [ACK]
63 29.039028    158.47.121.32 192.168.1.43  TCP  66 35314 > iso-tsap [ACK]
64 31.052553    192.168.1.43  158.47.121.32  MMS  179 unconfirmed-PDU
65 31.052653    158.47.121.32 192.168.1.43  TCP  66 35314 > iso-tsap [ACK]
```

Fig. 3. Traffic trace from the testbed (left) and the model (right)

Comparing the content of a MMS report frame obtained from the real application with a report frame from the ns-3 simulation with DCE, we see the same

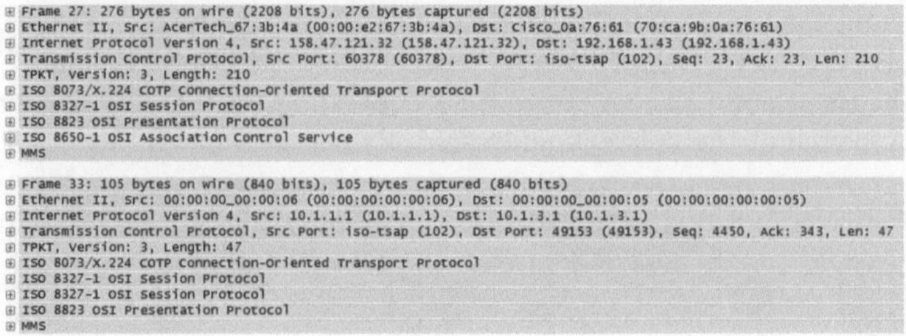

```
⊞ Frame 27: 276 bytes on wire (2208 bits), 276 bytes captured (2208 bits)
⊞ Ethernet II, Src: AcerTech_67:3b:4a (00:00:e2:67:3b:4a), Dst: Cisco_0a:76:61 (70:ca:9b:0a:76:61)
⊞ Internet Protocol Version 4, Src: 158.47.121.32 (158.47.121.32), Dst: 192.168.1.43 (192.168.1.43)
⊞ Transmission Control Protocol, Src Port: 60378 (60378), Dst Port: iso-tsap (102), Seq: 23, Ack: 23, Len: 210
⊞ TPKT, Version: 3, Length: 210
⊞ ISO 8073/X.224 COTP Connection-Oriented Transport Protocol
⊞ ISO 8327-1 OSI Session Protocol
⊞ ISO 8823 OSI Presentation Protocol
⊞ ISO 8650-1 OSI Association Control Service
⊞ MMS

⊞ Frame 33: 105 bytes on wire (840 bits), 105 bytes captured (840 bits)
⊞ Ethernet II, Src: 00:00:00_00:00:06 (00:00:00:00:00:06), Dst: 00:00:00_00:00:05 (00:00:00:00:00:05)
⊞ Internet Protocol Version 4, Src: 10.1.1.1 (10.1.1.1), Dst: 10.1.3.1 (10.1.3.1)
⊞ Transmission Control Protocol, Src Port: iso-tsap (102), Dst Port: 49153 (49153), Seq: 4450, Ack: 343, Len: 47
⊞ TPKT, Version: 3, Length: 47
⊞ ISO 8073/X.224 COTP Connection-Oriented Transport Protocol
⊞ ISO 8327-1 OSI Session Protocol
⊞ ISO 8327-1 OSI Session Protocol
⊞ ISO 8823 OSI Presentation Protocol
⊞ MMS
```

Fig. 4. MMS packet structure from the testbed (top) and the model (bottom)

structure (Fig. 5): in both case the report fields of the packet are exactly those specified by the IEC 61850 standard.

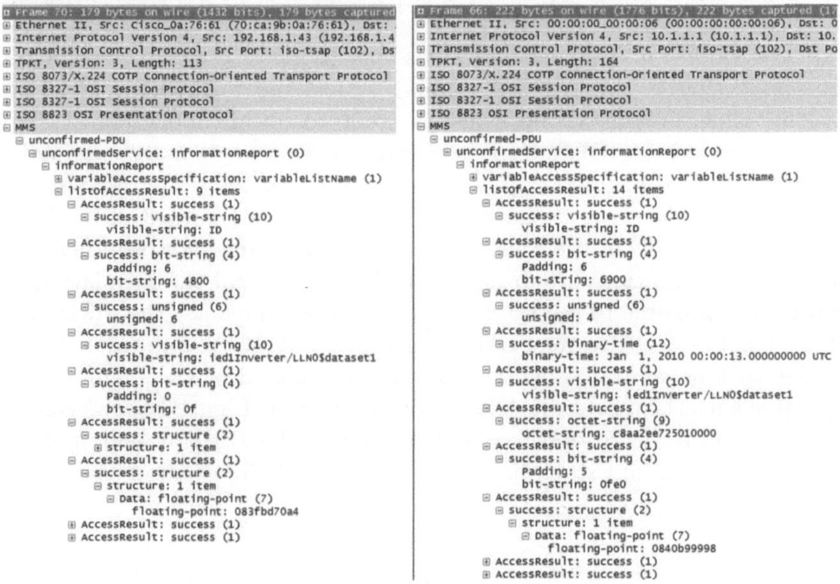

```
⊞ Frame 70: 179 bytes on wire (1432 bits), 179 bytes captured     ⊞ Frame 66: 222 bytes on wire (1776 bits), 222 bytes captured (1
⊞ Ethernet II, Src: Cisco_0a:76:61 (70:ca:9b:0a:76:61), Dst:       ⊞ Ethernet II, Src: 00:00:00_00:00:06 (00:00:00:00:00:06), Dst: 0
⊞ Internet Protocol Version 4, Src: 192.168.1.43 (192.168.1.4      ⊞ Internet Protocol Version 4, Src: 10.1.1.1 (10.1.1.1), Dst: 10.
⊞ Transmission Control Protocol, Src Port: iso-tsap (102), Ds      ⊞ Transmission Control Protocol, Src Port: iso-tsap (102), Dst Po
⊞ TPKT, Version: 3, Length: 113                                    ⊞ TPKT, Version: 3, Length: 164
⊞ ISO 8073/X.224 COTP Connection-Oriented Transport Protocol       ⊞ ISO 8073/X.224 COTP Connection-Oriented Transport Protocol
⊞ ISO 8327-1 OSI Session Protocol                                  ⊞ ISO 8327-1 OSI Session Protocol
⊞ ISO 8327-1 OSI Session Protocol                                  ⊞ ISO 8327-1 OSI Session Protocol
⊞ ISO 8823 OSI Presentation Protocol                               ⊞ ISO 8823 OSI Presentation Protocol
⊟ MMS                                                              ⊟ MMS
  ⊟ unconfirmed-PDU                                                  ⊟ unconfirmed-PDU
    ⊟ unconfirmedService: informationReport (0)                       ⊟ unconfirmedService: informationReport (0)
      ⊟ informationReport                                              ⊟ informationReport
        ⊞ variableAccessSpecification: variableListName (1)             ⊞ variableAccessSpecification: variableListName (1)
        ⊟ listOfAccessResult: 9 items                                  ⊟ listOfAccessResult: 14 items
          ⊟ AccessResult: success (1)                                    ⊟ AccessResult: success (1)
            ⊟ success: visible-string (10)                                ⊟ success: visible-string (10)
              visible-string: ID                                           visible-string: ID
          ⊟ AccessResult: success (1)                                    ⊟ AccessResult: success (1)
            ⊟ success: bit-string (4)                                     ⊟ success: bit-string (4)
              Padding: 6                                                   Padding: 6
              bit-string: 4800                                            bit-string: 6900
          ⊟ AccessResult: success (1)                                    ⊟ AccessResult: success (1)
            ⊟ success: unsigned (6)                                       ⊟ success: unsigned (6)
              unsigned: 6                                                  unsigned: 4
          ⊟ AccessResult: success (1)                                    ⊟ AccessResult: success (1)
            ⊟ success: visible-string (10)                                ⊟ success: binary-time (12)
              visible-string: ied1Inverter/LLN0$dataset1                   binary-time: Jan  1, 2010 00:00:13.000000000 UTC
          ⊟ AccessResult: success (1)                                    ⊟ AccessResult: success (1)
            ⊟ success: bit-string (4)                                     ⊟ success: visible-string (10)
              Padding: 0                                                   visible-string: ied1Inverter/LLN0$dataset1
              bit-string: 0f                                             ⊟ AccessResult: success (1)
          ⊟ AccessResult: success (1)                                     ⊟ success: octet-string (9)
            ⊟ success: structure (2)                                        octet-string: c8aa2ee725010000
              ⊞ structure: 1 item                                        ⊟ AccessResult: success (1)
          ⊟ AccessResult: success (1)                                     ⊟ success: bit-string (4)
            ⊟ success: structure (2)                                        Padding: 5
              ⊟ structure: 1 item                                          bit-string: 0fe0
                ⊟ Data: floating-point (7)                               ⊟ AccessResult: success (1)
                  floating-point: 083fbd70a4                              ⊟ success: structure (2)
          ⊞ AccessResult: success (1)                                      ⊟ structure: 1 item
          ⊞ AccessResult: success (1)                                        ⊟ Data: floating-point (7)
                                                                              floating-point: 0840b99998
                                                                       ⊞ AccessResult: success (1)
                                                                       ⊞ AccessResult: success (1)
```

Fig. 5. MMS packet content from the testbed (top) and the model (bottom)

4.2 Models for the Flooding Attack Scenarios

Starting from the legal communication model validated with the testbed traces, now we introduce a variable number of additional nodes representing one or more attackers connected to the DER HV/MV substation Wide Area Network as displayed in Fig. 6. Each attacker runs the flooding attack tool taken from the testbed that sends illegal packets with source one or more attackers (node 8

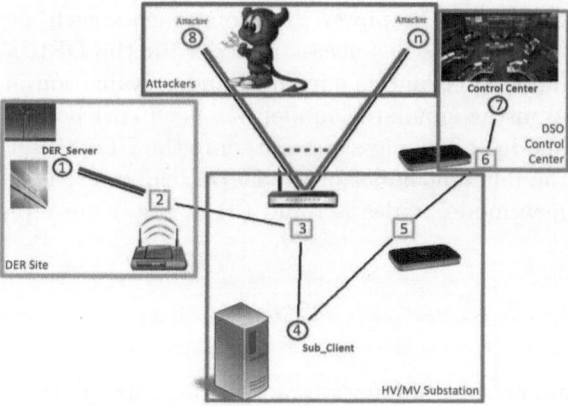

Fig. 6. Flooding attack

to n) and target the substation router (3), i.e. the same interface that receives the legal MMS packets from the DER.

In Fig. 7 the pcap trace is displayed where we see the MMS report and, after the beginning of the attack, the malicious UDP packets masqueraded as Syslog messages, a legal message type in the communications for ICT monitoring.

No.	Time	Source	Destination	Protocol	Length	Info
66	9.490018	10.1.1.1	10.1.3.1	MMS	222	unconfirmed-PDU
67	9.690041	10.1.3.1	10.1.1.1	TCP	64	49153 > iso-tsap [ACK] Seq=805 Ack=5497 Win=65535 Len=0
68	11.490018	10.1.1.1	10.1.3.1	MMS	222	unconfirmed-PDU
69	11.690041	10.1.3.1	10.1.1.1	TCP	64	49153 > iso-tsap [ACK] Seq=805 Ack=5661 Win=65535 Len=0
70	13.490018	10.1.1.1	10.1.3.1	MMS	222	unconfirmed-PDU
71	13.690041	10.1.3.1	10.1.1.1	TCP	64	49153 > iso-tsap [ACK] Seq=805 Ack=5825 Win=65535 Len=0
72	15.490018	10.1.1.1	10.1.3.1	MMS	222	unconfirmed-PDU
73	15.690041	10.1.3.1	10.1.1.1	TCP	64	49153 > iso-tsap [ACK] Seq=805 Ack=5989 Win=65535 Len=0
74	16.495988	00:00:00_00:00:00	Broadcast	ARP	64	who has 10.1.2.1? tell 10.1.2.3
75	16.495988	00:00:00_00:00:00	00:00:00_00:00:0	ARP	64	10.1.2.1 is at 00:00:00:00:00:05
76	16.496006	10.1.2.3	10.1.2.1	UDP	146	Source port: ddi-udp-1 Destination port: interwise
77	16.496019	10.1.2.3	10.1.2.1	UDP	146	Source port: ddi-udp-1 Destination port: interwise
78	16.496032	10.1.2.3	10.1.2.1	UDP	146	Source port: ddi-udp-1 Destination port: interwise
79	16.496077	10.1.2.3	10.1.2.1	Syslog	558	Attack UDP message from: NODE_4 to 10.1.2.1 : 514\0008\000\
80	16.496123	10.1.2.3	10.1.2.1	Syslog	558	Attack UDP message from: NODE_4 to 10.1.2.1 : 514\0008\000\
81	16.496168	10.1.2.3	10.1.2.1	Syslog	558	Attack UDP message from: NODE_4 to 10.1.2.1 : 514\0008\000\
82	16.496214	10.1.2.3	10.1.2.1	Syslog	558	Attack UDP message from: NODE_4 to 10.1.2.1 : 514\0008\000\
83	16.496260	10.1.2.3	10.1.2.1	Syslog	558	Attack UDP message from: NODE_4 to 10.1.2.1 : 514\0008\000\
84	16.496305	10.1.2.3	10.1.2.1	Syslog	558	Attack UDP message from: NODE_4 to 10.1.2.1 : 514\0008\000\

Fig. 7. Trace of attack scenario

Thanks to the parameters of the attack tool, we perform a sensitivity analysis of the effects of the attack on the MMS communication performance by varying the number of attackers, the packet rate and payload size of the illegal traffic.

4.3   Model of the LTE Technology

An important aspect that characterizes the communication between the primary substation and the DERs is the heterogeneity of the technologies deployed. The DER sites can be placed in areas where the wired connections are absent. Thanks to the new developments, the mobile technologies may represent a valid solution for the communication of the power grid components such as the DERs. The LTE technology may be used as access network for the DER and the HV/MV substation communications and is currently under evaluation in the lab testbed [15]. For this reason the simulated model has been enriched with the inclusion of the ns-3 LTE module. The messages outgoing the LTE router reach the destination through the different nodes of the LTE architecture included in the LTE module. In the new model nodes (2) and (3) in Fig. 1 are represented by LTE routers.

5   Results

This section shows how the simulation models presented in the previous sections can be deployed in order to define the attack scenarios for the experimental setup with the testbed. The simulated scenarios are selected considering the attack applications running on the testbed in order to investigate as the attack parameter configuration affects the legal communications. Moreover the strength of the simulation environment is that it allows studying larger scenarios than those feasible in the testbed, for example scaling up the number of connected DER.

5.1   Setup Parameters

In this subsection different flooding attack parameters are considered and the attack impact on the DER primary substation communication delay is shown. In the first analysis the focus is on evaluating the impact of the number of attackers on the MMS communication delay, in particular considering the transmission time taken by the DER measurements to reach the Primary substation. At time 1 the MMS client /server applications are activated and at time 8 the UDP flooding attack process starts. Figure 8 shows the delay values in normal and attack scenarios. It is possible to note how, varying the number of attackers, the effect on the legal communication changes. If only one attacker is involved, the communication delay has a deviation from the normal behavior, but the two parties are able to exchange messages. Increasing the number of the attackers the delay time is increasing until the point in time when the connectivity is lost (e.g. time 21 with 2 attackers). The connectivity loss occurs when the protocol timeouts expires and it is not possible to maintain the current session active. This happens earlier by increasing the number of attackers. Another important parameter to be taken in consideration for the setup of the testbed experiments is the frequency of the malicious packet emissions. In Fig. 9 the delay values considering different time between two consecutive packet emissions (10000, 1000 and 100 microseconds) are plotted. All the attack scenarios lead to a connection loss, but earliest with higher frequency. It is possible to note that in case of frequency set to 100 microseconds there is not a peak but an adjustment of the delay value. In the last set of experiments, the malicious packet size is taken as parameter under observation. In Fig. 10 the plot of the delay values in normal and

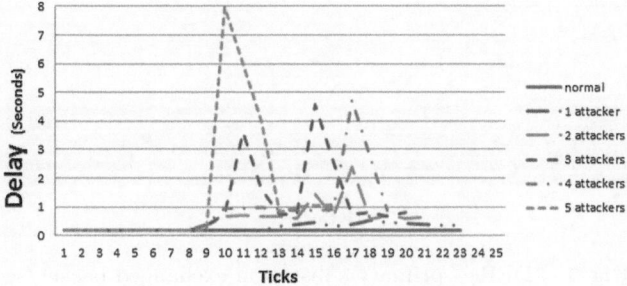

Fig. 8. Attack effect on delay changing the number of attackers

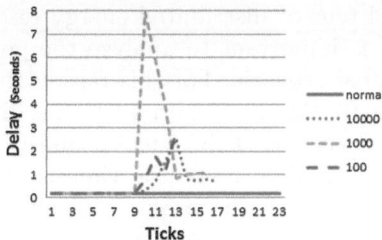

Fig. 9. Attack effect on delay changing the frequency

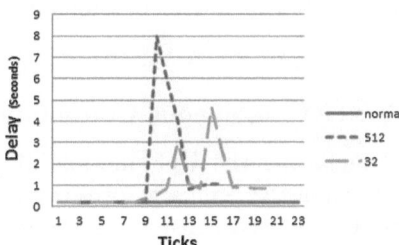

Fig. 10. Attack effect on delay changing the attack packet size

under attack conditions are shown. In particular two packet sizes are considered: 32 bytes and 512 bytes. The legal packet containing the DER measurements has a size around 200 bytes, so a smaller and a bigger malicious packet size have been taken in the simulations. If the packet size is double respect of the legal one it is possible to see an early connection loss, this happens also in case of the smaller attack packet size, but later with more dilated time.

5.2   Scaling up the Model Architecture

The simulation model can be deployed in order to explore the scenarios not addressable by the testbed experiments, for example scaling up the number of DER connected to grid. Indeed in order to satisfy the European targets of DER

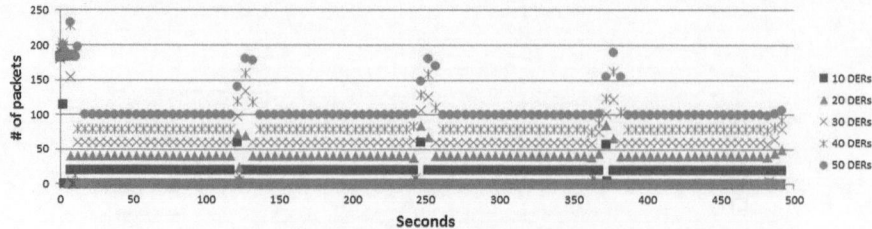

Fig. 11. DERs - primary substation exchanged packet/sec

penetration the full roll out of active grids requires that a single primary sub-station is able to control tens of distributed energy resources connected to the medium voltage grid. It is important to analyze the performance of communi-cation in such a distributed scenario. For this reason the simulation model has been extended including 10, 20, 30 40 and 50 DERs. In Fig. 11 the amount of packet/sec exchanged between the DERs and a primary substation is shown con-sidering different model size. The simulation time is of 500 s and it is possible to identify a first phase in which the profile of each DER is exchanged. This requires a large amount of exchanged packets and then each DER sends peri-odically, every 2 s, a MMS report containing its measurements. Some of these packets need to be retransmitted in order to reach the application in the pri-mary substation, and the number of retransmissions increases with the number of DERs.

5.3 Experiments with Heterogeneous Communication Technologies

The setup of the model including the LTE module allows analyzing the per-formance of the mobile technology versus the results obtained with the wired layout. In Fig. 12 the results achieved considering the round trip time (RTT) as indicator are plotted. From the simulation outcome it is possible to argue that the RTT values achieved deploying the LTE module are comparable with the base line communication, also considering a short plus delta resulting by the LTE model. In case of DER measurements (in Fig. 12 from packet 26 to the end of the simulation) the increase is of 17.757 ms.

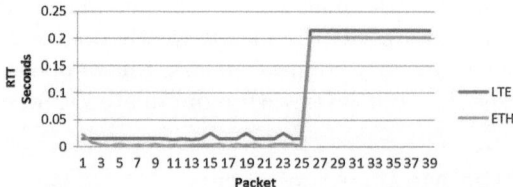

Fig. 12. RTT in wired vs wireless communications

6   Conclusion

The use of ICT network simulators for the assessment of the attack effects on smart grid communications is a promising approach for studying critical scenarios, whose results present synergies and interrelationships with experimental activities. However it still is a challenging research objective to achieve simulation results that are mostly aligned with the performance indicators measured in the real ICT architecture. The simulation tool has to incorporate into the model all the aspects of real architecture, i.e. the protocol aware information flows, the security countermeasures, the network topology and the malicious processes. In this paper the DER-substation information flow of a voltage control function for active grids has been modeled by a network simulator integrating a real client/server application implemented in a testbed setup. This approach allowed representing realistic traffic profiles in the simulation runs. The simulation outcomes have been validated with the testbed traces verifying a full alignment between the real and the simulated traffic profile. The results from the sensitivity analysis of the flooding attack effects by varying the attack parameters are discussed and used in order to identify the relevant scenarios for the testbed experiments. In the future work the effect of other attack tools will be evaluated with the simulator over a communication architecture enhanced with the security measures for the node authentication and the data encryption already implemented in the testbed, in compliance with the end-to-end security standard IEC 62351-3.

Acknowledgments. The work has been supported by the European FP7 project SmartC2Net (under grant agreement no. 318023). Further information is available at www.smartc2net.eu.

References

1. http://www.opnet.com
2. Mller, S.C., Georg, H., Rehtanz, C., Wietfeld, C.: Hybrid simulation of power systems and ICT for real-time applications. In: 3rd IEEE PES Innovative Smart Grid Technologies Europe (ISGT Europe), Berlin (2012)
3. Georg, H., Wietfeld, C., Mller, S.C., Rehtanz, C.: A HLA based simulator architecture for co-simulating ICT based power system control and protection systems. In: 3rd IEEE International Conference on Smart Grid Communications (2012)
4. http://web.scalable-networks.com/content/qualnet
5. Ullo, S.L., Vaccaro, A., Velotto, G.: Performance analysis of IEEE 802.15. 4 based sensor networks for smart grids communications. J. Electr. Eng. Theory Appl. **1**(3), 129–134 (2010)
6. http://www.omnetpp.org/
7. Mets, K., Verschueren, T., Turck, F.D., Develder, C.: Exploiting V2G to optimize residential energy consumption with electrical vehicle (dis) charging. In: First International Workshop on Smart Grid Modeling and Simulation (SGMS), pp. 7–12, Brussels, October 2011

8. http://nsnam.isi.edu/nsnam/index.php/Main_Page

9. NS3 network simulator. https://www.nsnam.org/

10. Jung, J.-S., Lim, K.-W., Kim, J.-B., Ko, Y.-B., Kim, Y., Lee, S.-Y.: Improving IEEE 802.11s wireless mesh networks for reliable routing in the smart grid infrastructure. 2nd Workshop on Smart Grid Communications, pp. 1–5, Kyoto (2011)

11. Godfrey, T., Mullen, S., Dugan, R.C., Rodine, C., Grith, D.W., Golmie, N.: Modeling smart grid applications with co-simulation. In: 1st IEEE International Conference on Smart Grid Communications, pp. 291–296, Gaithersburg, October 2010

12. Moneta, D., Mora, P., Belotti, M., Carlini, C.: Integrating larger RES share in distribution networks: advanced voltage control and its application on real MV networks in Integration of Renewables into the Distribution Grid, CIRED 2012 Workshop, Lisbon, May 2012

13. SmartC2Net European Project, Deliverable D1.1 SmartC2Net Use Cases, Preliminary Architecture and Business Drivers, September 2013. http://www.smartc2net.eu

14. Int. Standard IEC 61850-8-1 (ed. 2) Communication networks and systems in substations - Part 8–1: Specific Communication Service Mapping (SCSM) - Mappings to MMS (ISO 9506–1 and ISO 9506–2) and to ISO/IEC 8802–3, June 2011

15. SmartC2Net European Project, Deliverable D6.2 "Integrated test beds - Description", May 2015. http://www.smartc2net.eu

16. libIEC61850 open source libarary for IEC 61850. http://libiec61850.com/

17. RSE PCS-ResTest Lab (Power Control Systems Resilience Testing Laboratory). http://www.rse-web.it/laboratori.page?RSE_originalURI=/laboratori/laboratorio/10&RSE_manipulatePath=yes&country=eng

jOSEF: A Java-Based Open-Source Smart Meter Gateway Experimentation Framework

Michael Hoefling$^{(\boxtimes)}$, Florian Heimgaertner, Daniel Fuchs, and Michael Menth

Chair of Communication Networks, University of Tuebingen, Tuebingen, Germany
{hoefling,florian.heimgaertner,menth}@uni-tuebingen.de,
daniel.fuchs@student.uni-tuebingen.de

Abstract. Smart meter gateways are the core component of the advanced metering infrastructure in Germany, and provide a unified interface for metering data retrieval to third parties. Different standards and communication protocols exist for smart metering, ranging from transmission protocols to architectural recommendations. This work briefly presents the concept of the German BSI TR-03109 smart metering architecture, reviews implementations of smart metering protocols and architectures, and provides a *Java-based open-source smart meter gateway experimentation framework (jOSEF)*. The proposed framework combines and extends established protocol frameworks to provide a flexible tool for the validation of smart metering communication use cases involving smart meter gateways.

Keywords: Smart grid · Smart metering · Smart meter gateway · Advanced metering infrastructure · BSI TR-03109

1 Introduction

Electrical power distribution networks are undergoing major changes in operational procedures and monitoring, thereby evolving from passive to active networks [1,2]. One consequence of these changes is the introduction of *advanced metering infrastructures (AMIs)* in the distribution grid to provide automatic billing, and acquisition of network status data. *Smart meters (SMs)* are the basic component of AMIs, replacing traditional electricity, gas and heat meters in the long run. SMs measure energy consumption and production in private households, commerce, and the industry, and provide mechanisms for remote meter reading. *Smart meter gateways (SMGWs)* gather metering information from several SMs, and provide a unified interface for meter information retrieval to interested and legitimate *external market participants (EMPs)*.

The *Cyber-secure Data and Control Cloud for power grids* (C-DAX) project [3] is an FP7 project funded by the European Commission which aims to develop a cyber-secure communication middleware for smart grids, applying the publish/subscribe (pub/sub) paradigm to enable scalable, transparent, and secure end-to-end communication [4] between publishers and subscribers.

© Springer International Publishing Switzerland 2015
S. Gottwalt et al. (Eds.): EI 2015, LNCS 9424, pp. 165–176, 2015.
DOI: 10.1007/978-3-319-25876-8_14

Additional major advantages of the C-DAX architecture include resilient communication [5], inter-domain communication, and support for real-time applications [6]. This work was conducted as part of the C-DAX project to evaluate the suitability of the C-DAX architecture for smart metering.

The main contribution of this paper is a brief presentation of the SMGW-based smart metering architecture as defined in BSI TR-03109 [7], and a description of jOSEF, a Java-based open-source SMGW experimentation framework. While existing implementations allow the isolated simulation and evaluation of certain smart metering communication aspects, a framework providing the minimally necessary building blocks for a BSI TR-03109-compliant SMGW-based architecture has been missing in both literature and in practice. This work addresses this gap and actually allows to model an SMGW-based smart metering architecture utilizing open-source components. This work mainly focuses on the German AMI approach [7] but considers the Dutch AMI approach [8] for technical details that have not been defined for Germany yet.

This work is structured as follows. We review relevant protocols for smart metering in Sect. 2, and present the concept of the BSI TR-03109 smart metering architecture in Sect. 3. In Sect. 4, we describe existing implementations and discuss their suitability for the development of a BSI TR-03109-compliant SMGW. Section 5 describes jOSEF in detail and Sect. 6 illustrates its functionality with selected communication scenarios. Section 7 concludes this work.

2   Related Work

The DLMS/COSEM suite is a set of standards for the exchange of energy meter data, comprising of *DLMS (device language message specification)* [9] as an application layer protocol for communication with metering devices, and *COSEM (companion standard for energy metering)* [10] as a system for object-oriented modeling of energy metering equipment. DLMS/COSEM uses the *object identification system (OBIS)* [11] to identify data objects in energy metering systems, and *COSEM services* enable clients to query specific attributes of objects, assign values to attributes of objects, or execute methods of objects.

SML (smart message language) [12] is a message-oriented protocol for communication with SMs. The SML application protocol defines *SML files* consisting of one or multiple *SML messages*. An *SML message* can be either a request or a response. SMs act as servers, receiving SML files from clients, and processing the contained SML messages in order of reception. Starting with version 1.04, SML supports COSEM services, i.e., the COSEM object model can be used with the SML application protocol. Currently, SML is not widely used outside Germany. However, international use of SML is expected to increase if plans to adopt SML as part of the DLMS/COSEM suite [13] are successful.

M-Bus is a protocol suite for communication with SMs. M-Bus is defined in the European standard EN 13757 which comprises data model [14], application layer [15], and both wired [16] and wireless [17] specifications for the physical layer. The Open Metering System (OMS) [18,19] is a smart metering communication architecture based on M-Bus. OMS proposes several modifications to the

M-Bus protocols, and adds an optional authentication and fragmentation layer to the M-Bus protocol stack.

The *Dutch SM requirements (DSMR)* [8] are a joint specification of the Dutch grid operators. DSMR is based on DLMS/COSEM and M-Bus, and defines a data model for SMs including corresponding OBIS codes. We use selected parts of DSMR to fill the technical gaps of the German BSI TR-03109 for our framework.

3 Smart Meter Gateways: A Communication Topology for Smart Metering

Smart meter gateways (SMGWs) are the central communication components in the future smart metering infrastructure in Germany [7,20]. The two most important functionalities of SMGWs are (1) gathering of metering data from SMs, and (2) providing a unified interface for metering data retrieval to interested and legitimate EMPs.

In general, the SMGW mediates between three networks, as shown in Fig. 1: the *local metrological network (LMN)*, the *home area network (HAN)*, and the *wide area network (WAN)*. The LMN connects SMs to the SMGW only. The HAN connects end consumers, service technicians, and *controllable local systems (CLSes)* to the SMGW, e.g., electric vehicles, photo-voltaic panels, and remote-controllable heating and air conditions. The WAN connects administrators and EMPs to the SMGW, e.g., distribution grid operators, metering point operators, and suppliers of electric energy.

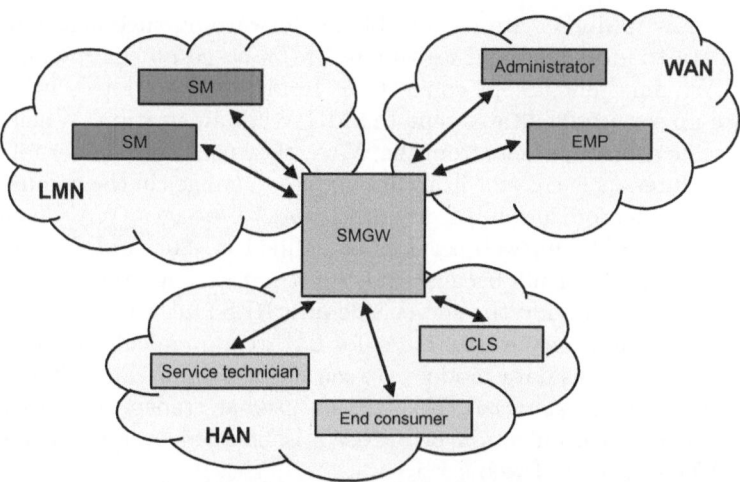

Fig. 1. System boundaries of the SMGW architecture according to [20]. The SMGW mediates between LMN, HAN, and WAN.

3.1   Functionalities and Communication

The functionalities and the used communication protocols of SMGWs can be differentiated by the networks they mediate between.

In the LMN, SMGWs are responsible for gathering metering data from SMs according to metering profiles, time-stamping the measurements based on an externally synchronized time source, tariffing, and finally storing the time-stamped, tariffed metering data for further dissemination to EMPs. SMGWs support bidirectional and unidirectional communication with SMs. Bidirectional communication involves interactive communication between SMGWs and SMs to poll for metering data or to manage SMs. Unidirectional communication stands for unsolicited metering data dissemination from SMs to SMGWs. Generally, COSEM [10] with OBIS [11] codes are used as data model between SMs and SMGWs. Depending on the underlying physical layer, M-Bus [14–17] or SML [12] is used as transport protocol.

In the HAN, SMGWs provide read-only access to their internally stored metering data and status messages to end consumers. SMGWs can support several end consumers facilitating multi-client operation, e.g., in an environment involving many SMs and many households. Service technicians must only access status messages of SMGWs. SMGWs relay control messages between CLSes and EMPs as configured by administrators. BSI TR-03109 [20] does not specify protocols between SMGWs and potential HAN communication partners but security mechanisms to be used, e.g., secure transport layer communication, and mandatory authentication of clients against the SMGW. Essentially, any IP-based protocol may be used between SMGWs and HAN entities, e.g., end consumers or service technicians.

In the WAN, SMGWs are responsible for forwarding their internally stored metering data to interested and legitimate EMPs based on communication profiles. SMGWs must not accept connections from the WAN for security reasons but a wake-up service facilitates remote SMGW administration. When SMGWs receive specific control packets from the WAN, they contact an external administrator for maintenance, e.g., for firmware updates, changes in the communication profiles, time synchronization, or access to status messages. WAN communication is based on RESTful web services as defined in [20], and SMGWs act as RESTful web service clients because they must not accept connections from the WAN. EMPs must provide the server side of a RESTful web service according to the interface definitions in [20,21]. As for LMN communication, COSEM with OBIS codes are used as data model between SMGWs and EMPs but XML and *cryptographic message syntax (CMS)* [22] are used as transport protocol on top of REST. Time synchronization of SMGWs is handled over the *network time protocol (NTP)* instead of web services.

3.2   Security

The BSI *SMGW protection profile (SMGW-PP)* [23] requires all LMN, HAN and WAN communication to be secured by *transport layer security (TLS)* in combination with a public-key infrastructure [24,25]. WAN communication is further protected by CMS between SMGWs and EMPs. SMGWs are equipped with a security module which provides cryptographic functions, e.g., generation and secure storage of encryption keys, and verification of digital certificates. The security module is realized as a smart card. Further information on the security module and its requirements can be found in [26–28].

These security requirements limit the suitability of the C-DAX middleware for smart metering in Germany because C-DAX provides its own strong security mechanisms [4] but does not support TLS between communication partners without modification. However, if the BSI security regulations would permit replacing TLS by other security mechanisms with the same level of security, C-DAX may be used as communication middleware for HAN communication, e.g., between SMGWs and EMPs. In that case, C-DAX' pub/sub mechanisms would allow scalable, secure and resilient dissemination of tariff information or firmware updates to all SMGWs, or transparent and secure remote control of a customers CLSes.

4   Existing Implementations

In this section, we review selected open-source implementations and discuss their suitability for the development of a BSI TR-03109 compliant SMGW.

OpenMUC [29] is an open-source implementation of a *multi utility communication controller (MUC)* developed at Fraunhofer ISE. OpenMUC is implemented in Java and licensed under the terms of the GNU General Public License (GPL). The core component of OpenMUC is the data manager which interfaces to optional components like data server, logger, protocol drivers, and custom applications. The OpenMUC framework provides the functionality specified in a previous draft standard for a German AMI. While BSI TR-03109 requires the use of XML for the RESTful web service, OpenMUC uses the JSON format. Additionally, the URI hierarchy used by OpenMUC differs from the BSI specification, and the protocol drivers do not satisfy the minimum requirements. Implementing a BSI TR-03109 compliant SMGW based on OpenMUC would require major changes to the OpenMUC code. However, the OpenMUC framework also includes protocol libraries that can be used independently, e.g., jDLMS and jSML. We discuss those in the following subsections.

jDLMS [30] is a Java implementation of the DLMS/COSEM protocol available under the terms of the GNU Lesser General Public License (LGPL). jDLMS supports the DLMS/COSEM application layer protocol over serial lines using HDLC, or over TCP or UDP. As the current version 0.9.0 only implements the client side of the DLMS/COSEM protocol, and does not include the COSEM object model, jDLMS is not suitable for implementing a SMGW according to BSI TR-03109.

jSML [31] is a Java implementation of SML available under the terms of the LGPL. jSML supports SML communication over TCP/IP or over serial line. HDLC support is currently unavailable. jSML implements the SML message format and encoding, the SML data types, and the SML transport layer version 1. The current version 1.0.17 is based on SML 1.03, i.e., jSML does not yet support COSEM services. For implementing a SMGW according to BSI TR-03109, jSML needs to be extended to support the current SML version.

Open gateway energy management (OGEMA) [32] is an OpenMUC-based software platform for building automation and load management. OGEMA is implemented in Java and licensed under the terms of the GPL. Since OGEMA is based on OpenMUC and focuses rather on the HAN side than on the SMGW, it is not suitable for implementing a SMGW according to BSI TR-03109.

Gurux [33] is a collection of smart metering software components developed by Gurux Ltd. Gurux code contains implementations in C#, C++, Java, and Delphi and is licensed under the terms of the GPL. Gurux supports DLMS/COSEM, Modbus, and M-Bus. However, the COSEM object model is not separated from the DLMS/COSEM application protocol in the Gurux code. Using the COSEM object model with other application protocols as required by BSI TR-03109, would require major modifications to the code.

5   jOSEF: A Java-Based Open-Source SMGW Experimentation Framework

We now present jOSEF, a Java-based open-source SMGW experimentation framework. jOSEF is licensed under the terms of the GPL version 2 or later. We describe its architecture, specify its operation, and discuss the deviation from BSI TR-03109. The implementation utilizes the jSML library [31] for LMN communication that was extended as part of this work to support SML version 1.04 [12]. Additionally, we used the COSEM implementation of Gurux [33] as a blueprint to implement DSMR's COSEM object model. We used DSMR's COSEM object model because BSI TR-03109 has not defined a companion standard for its COSEM object model yet.

5.1   Components

The framework comprises three main components: a minimal SMGW, an SM simulator, and a simple EMP. The *minimal SMGW* represents an SMGW which provides the minimally necessary functionalities to control SMs and to send meter data to EMPs. It is equipped with a GUI for configuration and operation, and allows several SMs to be connected, as shown in Fig. 2. The *SM simulator* represents an SM and can be configured with standard load profiles for energy generation and consumption to simulate SM behavior. It is controlled over a CLI, as shown in Fig. 3. The *simple EMP* provides an BSI TR-03109-compliant RESTful web service towards the SMGW and acts as a data sink for meter data. It can be accessed using any HTTP client, e.g., a web browser.

5.2   Meter Data Retrieval

When the SMGW wants to retrieve meter data from an SM, it first sends an SML message to the SM requesting all internal COSEM object IDs to discover

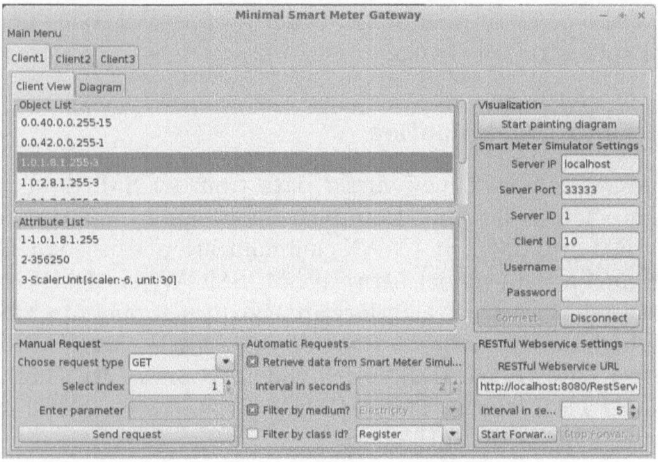

Fig. 2. Screenshot of the SMGW GUI of jOSEF. Client 1, 2, and 3 correspond to three different SMs that are connected to the SMGW, and the attribute list pane shows a detailed view of attribute 1.0.1.8.1.255 (electricity consumption in Wh) of client 1.

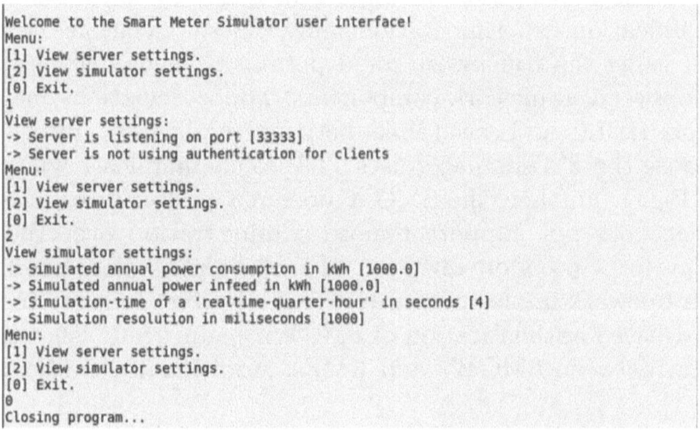

Fig. 3. Console log of the SM simulator CLI. The CLI allows the user to view its configuration. The configuration may only be changed on SM simulator startup via a configuration file.

the SM's internal data model. The returned list of COSEM object IDs is then used by the SMGW to build the actual meter data retrieval request by filtering for metering object IDs based on OBIS codes. The SMGW generates a new SML message containing explicit requests for details on the metering object IDs, and sends the message to the SM. The SM returns the actual metering objects to the SMGW that can perform further processing on the data, e.g., time stamping, tariffing, buffering, or dissemination to EMPs. The SMGW re-sends the meter data request message to the SM to receive new meter data;

rediscovery of the SM's internal data model by the SMGW is only necessary when the SM configuration changes.

5.3 Meter Data Dissemination

When the SMGW retrieved new meter data from an SM, data conversion is necessary before actual dissemination to EMPs because LMN communication is based on COSEM over SML, and WAN communication is based on COSEM over XML. The common data model between SM, SMGW and EMP is COSEM so that data conversion works straightforward, i.e., a mapping of COSEM objects to XML is defined in [21]. After conversion to COSEM over XML, the SMGW sends the meter data to the EMP using the appropriate RESTful web service endpoint and HTTP methods. The EMP stores the received meter data and can perform further processing on the data.

5.4 Limitations

Our current framework implementation deviates from BSI TR-03109 in some minor points which we consider not important if the framework is used for laboratory communication experimentations only. These deviations need to be considered when using the framework for experiments involving insecure network connections between framework components. Minor deviations include that we do not support HDLC and serial links between the SMGW and the SM at the moment because the SM simulator uses TCP to communicate with the SMGW, as shown in Fig. 4. Further, the SMGW does not perform tariffing on received meter data, it does not support remote administration over the WAN, and it does not perform pseudonymization of meter data before sending them to EMPs. The framework implements only a limited subset security functionalities, e.g., password-based authentication of SMGWs against SMs is available, but no authentication between SMGWs and EMPs. Additionally, we do not use TLS

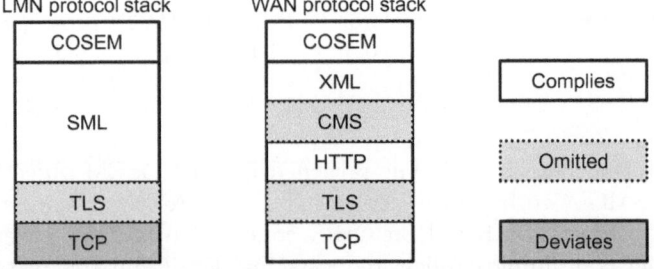

Fig. 4. Deviations of the LMN and WAN protocol stack of the current jOSEF implementation from BSI TR-03109. Currently omitted protocol layers are shown as gray, dashed boxes, and deviating protocol layers are shown as yellow, solid boxes (Color figure online).

for LMN and WAN communication, and we do not use CMS to further secure WAN communication, as shown in Fig. 4.

6 Illustration

We now illustrate the functionality of the proposed framework by experimentation. The setup of the experiment is described first, followed by experimental results from traffic experiments. Our results show that our framework enables easy modeling of typical smart metering topologies and communication patterns.

6.1 Experiment Setup and Methodology

To illustrate the functionality of the proposed framework, we created a simple dumbbell-like topology with one SM on the left side, one SMGW in the middle,

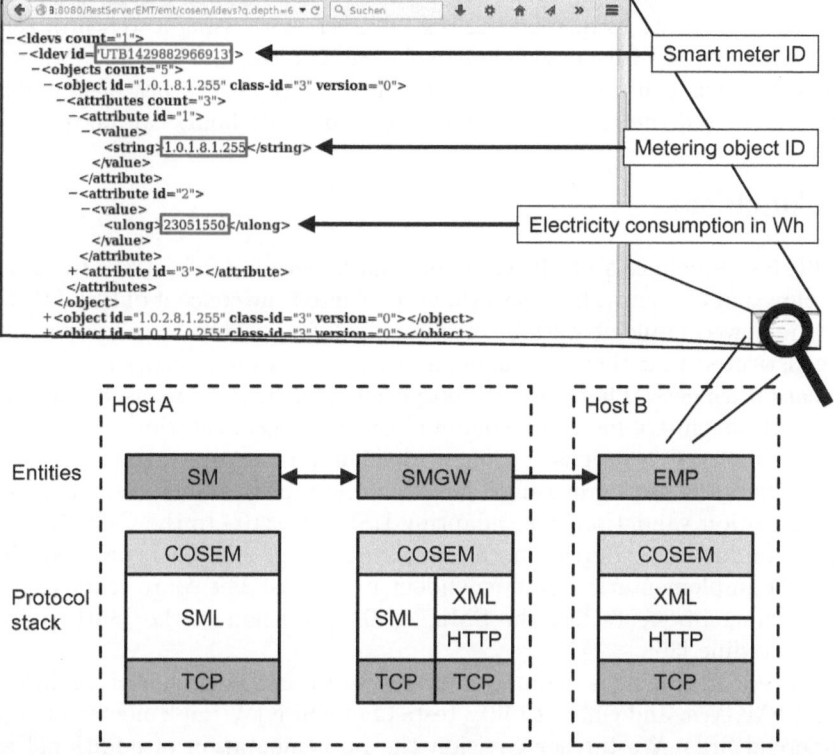

Fig. 5. Illustration of basic LMN and WAN communication including involved entities and protocol layers. A SMGW requests meter data from a SM, translates from COSEM over SML to COSEM over XML, and forwards the meter data to an EMP. The screenshot on the top shows what the XML structure of meter data looks like at the EMP.

and one EMP on the right side, as shown in Fig. 5. The SM was simulated by an SM simulator instance, configured to use a H0 load profile [34] and a EV0 generation profile [35]. The SMGW was configured to actively poll its associated SM every 2 s for new meter readings, and to forward all internally buffered metering data unsolicited to the EMP every 5 s. The EMP buffered meter readings from the SMGW, queryable via a RESTful web service interface. We deployed our setup on two end-hosts connected via a 100 Mbit/s Ethernet link.

6.2 Basic LMN and WAN Communication

First, the SM simulator, the SMGW and the EMP are started. The SMGW contacts the SM, queries for the SM's internal object list, and then subsequently polls for electricity objects only. When the SMGW receives the first meter data from the SM over SML, it translates from COSEM over SML to COSEM over XML, and starts forwarding the meter data to the EMP using HTTP PUT. The EMP stores the received meter data and provides access to it over a RESTful web service. The screenshot on top of Fig. 5 shows what meter data looks like at the EMP when the EMP is queried via HTTP GET, e.g., using a web browser. We can see that the SM with device id UTB1429882966913 consists of five electricity objects. For better illustration, only object 1.0.1.8.1.255 has been expanded in the figure, and the electricity consumption in Watt hours can be read.

7 Conclusion

BSI TR-03109 defines an SMGW-based smart metering infrastructure as it will be deployed in Germany. In this work, we presented the concept of BSI TR-03109, briefly reviewed implementations of smart metering protocols and architectures, and constituted that they only allow limited evaluation of smart metering communication aspects. Therefore, we proposed jOSEF, a Java-based open-source framework for smart metering communication experimentation, and evaluated its functionality. Our proposed framework combines and extends established protocol frameworks, thus providing a flexible tool for SMGW-based smart metering communication validation, e.g., adapting BSI TR-03109 to the C-DAX communication middleware. Furthermore, our extension of the jSML protocol library allows the implementation of independent programs. The source code of jOSEF and its subcomponents like the SML v.1.04 extension of the jSML library is available online [36].

Future work includes compatibility tests of our SM simulator against commercial SMGWs, and compatibility tests of our SMGW implementation against commercial SMs. We further envision the implementation of additional smart metering use cases such as SM management. Additionally, security mechanisms such as TLS support shall be included in the existing implementation.

Acknowledgement. The research leading to these results has received funding from the European Community's Seventh Framework Programme FP7-ICT-2011-8 under grant agreement n° 318708 (C-DAX). The authors alone are responsible for the content of this paper.

The authors thank Roelof Klein, and Martijn Kammerling, both working at Alliander, for valuable input and stimulating discussions.

References

1. CIGRE Working Group C6.11: Development and Operation of Active Distribution Networks, April 2011
2. Heydt, G.T.: The next generation of power distribution systems. IEEE Trans. Smart Grid **1**(3), 225–235 (2010)
3. C-DAX Consortium: Cyber-secure Data and Control Cloud for Power Grids. http://www.cdax.eu/. Accessed 28 Sept 2015
4. Heimgaertner, F., Hoefling, M., Vieira, B., Poll, E., Menth, M.: A security architecture for the publish/subscribe C-DAX middleware. In: Workshop on Security and Privacy for Internet of Things and Cyber-Physical Systems (IoT/CPS-Security) in Conjunction with IEEE International Conference on Communications (ICC), London, UK, June 2015
5. Hoefling, M., Heimgaertner, F., Menth, M., Katsaros, K.V., Romano, P., Zanni, L., Kamel, G.: Enabling resilient smart grid communication over the information-centric C-DAX middleware. In: ITG/GI International Conference on Networked Systems (NetSys), Cottbus, Germany, March 2015
6. Chai, W.K., Wang, N., Katsaros, K.V., Kamel, G., Melis, S., Hoefling, M., Vieira, B., Romano, P., Sarri, S., Tesfay, T., Yang, B., Heimgaertner, F., Pignati, M., Paolone, M., Menth, M., Pavlou, G., Poll, E., Mampaey, M., Bontius, H., Develder, C.: An information-centric communication infrastructure for real-time state estimation of active distribution networks. IEEE Trans. Smart Grid **6**(4) (2015)
7. Bundesamt für Sicherheit in der Informationstechnik: Technische Richtlinie BSI TR-03109
8. Netbeheer Nederland: Dutch Smart Meter Requirements: P1 Companion Standard. DSMR Version 5.0
9. International Electrotechnical Comission: Electricity Metering Data Exchange - The DLMS/COSEM Suite - Part 5–3: DLMS/COSEM Application Layer. IEC 62056-5-3 ed1.0 (2013)
10. International Electrotechnical Comission: Electricity Metering Data Exchange - The DLMS/COSEM Suite - Part 6–2: COSEM Interface Classes. IEC 62056-6-2 ed1.0 (2013)
11. International Electrotechnical Comission: Electricity Metering Data Exchange - The DLMS/COSEM Suite - Part 6–1: Object Identification System (OBIS). IEC 62056-6-1 ed1.0 (2013)
12. Bundesamt für Sicherheit in der Informationstechnik: BSI TR-03109-1 Anlage IV: Feinspezifikation Drahtgebundene LMN-Schnittstelle, Teil b: SML - Smart Message Language. SML Version 1.04
13. International Electrotechnical Comission: Electricity Metering Data Exchange - Part 5-3-8 Smart Message Language SML. IEC 62056-5-3-8 (future standard)
14. European Committee for Standardization: Communication Systems for and Remote Reading of Meters - Part 1: Data Exchange. EN 13757–1:2015–01 (2015)
15. European Committee for Standardization: Communication Systems for and Remote Reading of Meters - Part 4: Wireless Meter Readout. EN 13757–4:2014–02 (2014)

16. European Committee for Standardization: Communication Systems for and Remote Reading of Meters - Part 2: Physical and Link Layer. EN 13757-2:2004 (2004)
17. European Committee for Standardization: Communication Systems for and Remote Reading of Meters - Part 3: Dedicated Application Layer. EN 13757-3:2013-08 (2013)
18. OMS Group: Open Metering System Specification, Volume 1: General Part. OMS Spec Vol1 1.4.0 (2011)
19. OMS Group: Open Metering System Specification, Volume 2: Primary Communication, Version 4.0.2. OMS Spec Vol2 4.0.2 (2014)
20. Bundesamt für Sicherheit in der Informationstechnik: Anforderungen an die Interoperabilität der Kommunikationseinheit eines intelligenten Messsystems, Technische Richtlinie BSI TR-03109-1, Version 1.0
21. Bundesamt für Sicherheit in der Informationstechnik: BSI TR-03109-1 Anlage II: COSEM/HTTP Webservices
22. Bundesamt für Sicherheit in der Informationstechnik: BSI TR-03109-1 Anlage I: CMS-Datenformat für die Inhaltsdatenverschlüsselung und -signatur
23. Bundesamt für Sicherheit in der Informationstechnik: Schutzprofil für die Kommunikationseinheit eines intelligenten Messsystems für Stoff- und Energiemengen, Version 1.3. BSI SMGW-PP 1.3
24. Bundesamt für Sicherheit in der Informationstechnik: Kryptographische Vorgaben für die Infrastruktur von intelligenten Messsystemen, Technische Richtlinie BSI TR-03109-3, Version 1.1
25. Bundesamt für Sicherheit in der Informationstechnik: Public Key Infrastruktur für Smart Meter Gateways, Technische Richtlinie BSI TR-03109-4, Version 1.0
26. Bundesamt für Sicherheit in der Informationstechnik: Smart Meter Gateway - Anforderungen an die Funktionalität und Interoperabilität des Sicherheitsmoduls, Technische Richtlinie BSI TR-03109-2, Version 1.1
27. Bundesamt für Sicherheit in der Informationstechnik: Kryptographische Vorgaben für Projekte der Bundesregierung, Teil 3 - Intelligente Messsysteme, Technische Richtlinie BSI TR-03116-3
28. Bundesamt für Sicherheit in der Informationstechnik: Schutzprofil für das Sicherheitsmodul der Kommunikationseinheit eines intelligenten Messsystems für Stoff- und Energiemengen, Version 1.02. BSI SecMod-PP 1.02
29. Feuerhahn, S., Zillgith, M., Becker, R., Wittwer, C.: Implementation of an Open Smart Metering Reference Platform - OpenMUC. In: ETG-Kongress (2009)
30. Mueller-Bier, K.: jDLMS. http://www.openmuc.org/index.php?id=42. Accessed 28 Sept 2015
31. Feuerhahn, S., Buehrer, M.: jSML. http://www.openmuc.org/index.php?id=63. Accessed 28 Sept 2015
32. Fraunhofer IWES: Open Gateway Energy MAnagement - OGEMA. http://www.ogema.org/. Accessed 28 Sept 2015
33. Gurux Ltd.: Gurux Open Source Device Communication. http://www.gurux.fi/. Accessed 28 Sept 2015
34. KommEnergie: Lastprofile von KommEnergie. http://www.kommenergie.de/?id=140. Accessed 14 August 2015
35. Stadtwerke Emmendingen: Lastprofile der Stadtwerke Emmendingen. https://swe-emmendingen.de/netz/strom-netz/lastprofile/. Accessed 14 August 2015
36. jOSEF: A Java-Based Open-Source Smart Meter Gateway Experimentation Framework. http://kn.inf.uni-tuebingen.de/software/josef. Accessed 14 August 2015

Modeling and Simulation

OpenGridMap: An Open Platform for Inferring Power Grids with Crowdsourced Data

José Rivera[✉], Christoph Goebel, David Sardari, and Hans-Arno Jacobsen

Department of Computer Science, Technische Universität München (TUM),
Munich, Germany
j.rivera@tum.de

Abstract. The energy transition requires profound changes to the power grid, both on the transmission and distribution level. The ability to assess the impact of these changes, e.g., the integration of more solar power or electric mobility, requires data and tools that only exist partially today. The goal of this paper is to introduce OpenGridMap, a new project with the goal of creating an open platform for inferring realistic power grids based on actual data. Our vision is to provide a tool to researchers and practitioners that is able to produce realistic input data for simulation studies. OpenGridMap will support the entire process from data collection to formatting grid data for various purposes. We explore innovative ways to capture data and produce power grid approximations, e.g., using smartphone apps, expert classification, existing map APIs, and graph inference algorithms. The latest developments of the project can be found at opengridmap.com.

Keywords: Power grids · Power distribution · Geographic information systems · Crowdsourcing

1 Introduction

In many countries around the world, governments have committed to ambitious sustainability goals. In many cases, these goals include a transition to renewable power generation, mostly from wind and solar. Apart from being intermittent, wind and solar power generation is distributed and therefore supplies electricity at lower voltage. In Germany, over 90 % of the installed renewable energy capacity is connected at the distribution level [12]. In addition to the integration of variable renewables, electricity demand will change substantially, as well. On the one hand, further electrification is well underway, e.g., in the form of electric mobility and heat pumps. On the other hand, various initiatives are trying to reduce the energy consumption of traditional electric loads. Thus, even without actively controlling demand, the current developments will radically change the way power flows in the existing power grids. Several efforts to overcome these challenges using computer science techniques are currently beeing investigated as part of the Energy Informatics research field.

The current power grid infrastructure, in particular on the distribution level, was designed with certain assumptions in mind that will not hold in the future.

© Springer International Publishing Switzerland 2015
S. Gottwalt et al. (Eds.): EI 2015, LNCS 9424, pp. 179–191, 2015.
DOI: 10.1007/978-3-319-25876-8_15

Many recent studies have shown that this could lead to stability issues [12, 28, 29]. However, the level at which these important issues can be analyzed given todays data availability and analysis methods is modest. In fact, most available studies are based on such extreme simplifications that practical results cannot be expected. Academic studies are often carried out using standardized test feeders [12], not actual ones. Some distribution grid operators own digitized distribution grid data. However, in most cases, this data is either not digitized, or not available at all. A recent survey in the US revealed that 23 % of the electric distribution utility companies have incomplete data of their networks and over 30 % have outdated data [13].

The obvious solution to the problem of data availability is to actually collect the data and store it for further use, be it scenario-based simulation, or to enable new Smart Grid control approaches. Unfortunately, the traditional method for collecting this data is difficult, expensive, and intrusive [13]. The purpose of OpenGridMap is to overcome this problem using computer science techniques, including mobile app based crowdsourcing of grid data, machine learning based classification of visible grid components, and rule-based inference of invisible components. This paper presents a description and the latest results of the OpenStreetMap project.

Our paper is structured as follows: Sect. 2 offers an overview of the technical landscape for geographical power grid datasets, geographical information systems and crowdsourcing. The features of OpenGridMap are described in Sect. 3. The current components of OpenGridMap are presented in Sect. 4. Finally, Sect. 5 provides an outlook on the future project activities.

2   Technical Landscape

The digitalization of power grid data and the use of Geographical Information Systems (GIS) to display and analyze this data is not a new concept. Crowdsourcing is also an established concept. In the following, we review the technical landscape of this topics and explain how OpenGridMap builds on previous result to offer a solution to power grid inference based on crowdsourced data.

2.1   Availability of Power Grid Data

To simulate power flows, a identification and mapping of power grid infrastructure to geographic coordinates is necessary precondition. Many of the operational and business decisions of utilities rely on an accurate estimate of the state of their infrastructure. While this data is relatively well known on the level of the transmission grid, it is often not accurate and is sometimes even unknown on the sub-transmission and distribution level. As we already pointed out, a recent survey in the US revealed that 23 % of the electric distribution utility companies have incomplete data of their networks, and over 30 % have outdated data [13]. However, even this data is not openly available to researchers. Usually, the access to actual power grid data is very restricted. The main argument for these restrictions is security, i.e., the risk of someone sabotaging or attacking the power grid infrastructure. With these restrictions in place, researchers are forced to work with fabricated test cases. For distribution networks several such test cases exist, in particular the IEEE radial distribution test feeders [22], the CIGRE test feeders [27], and the PNNL feeders [26]. Although these test cases are often used and have certainly contributed to the reproducibility, and thus the comparability, of research results, it is important to point out the they are still invented and thus not usable in practice. In reality, power grids can be highly individual, e.g., due to geographic features, utility strategies, regional legislation, or simply coincidence. Thus, to establish a realistic grid data base for practical applications, we have to find new ways to collect actual grid data, even without the help of utilities.

There are several projects making an effort to offer realistic geographic power grid data. For instance, a publicly funded initiative in Bavaria, Energie-Atlas Bayern [3], offers a complete GIS energy solution for the state of Bavaria in Germany. The project EnergyMap [4] offers geographical information of several renewable energy generators for Germany. Furthermore, the German DENA reports [12] use geographical information of several utilities in Germany to study the future distribution grid expansions requirements. The OpenGridMap project differs from the aforementioned projects in that it aims to develop a novel system for inferring the power grid infrastructure at the level of distribution grids, and freely providing it in the required format for detailed computer simulations.

2.2 Geographic Information Systems

Geographic Information Systems (GIS) are systems designed to capture, store, manipulate, analyze, manage, and visualize all types of geographical data, in particular built infrastructure, such as streets and buildings.

A popular commercial software solution for integrating maps with application-specific data is ArcGIS [1]. ArcGIS is a platform that provides ready-to-use maps, apps, and templates to analyze geographical data. A free and open-source alternative to ArcGIS is Q-GIS [6]. Nonetheless, both platforms require a lot of customization in order to function as a power grid GIS. In contrast, OpenGridMap will be tailored for this specific purpose.

The most important asset of a GIS is its data. There are several freely available services that offer geographical data. Probably the best known is Google Maps [15]. However, after a certain number of requests has been reached, the data service of Google Maps is no longer available without charge. Also there are several restrictions for the API [7]: (i) no unauthorized copying, modification, creation of derivative works, or display of the content, (ii) no pre-fetching, caching, or storage of content, (iii) no mass downloads or bulk feeds of content. These restrictions make Google Maps unattractive for OpenGridMap.

An alternative to Google Maps is OpenStreetMap [17]. OpenStreetMap is a project that creates and distributes free worldwide geographic data. The project was initiated because most maps we think of as free actually have legal or technical restrictions on their use, precluding people from using them in creative, productive, or unexpected ways. The advantage of the OpenStreetMap is that it relies on a community to crowdsource its data. This allows it to have a huge database of information that is not available anywhere else, for free.

The use of GIS is not new to the power industry. However, not all utilities have access to a comprehensive GIS. In fact, according to the authors of [23], no utility currently has complete and up to date data on their system, let alone tools to manage it. After all, the introduction and maintenance of a GIS comes at a significant cost. There have been some efforts to develop freely available open source tools to handle geographic power systems data [9]. However, such efforts will not be successful without an efficient, scalable approach for collecting and managing this data. OpenGridMap sets out to tackle this challenge by relying on crowdsourcing and automatic inference.

2.3   Crowdsourcing Geographical Data

The collection of detailed geographical data is a time-consuming and complicated task. Crowdsourcing has proven to be an efficient method for the rapid and cost-effective collection of such data [18]. The most prominent example is OpenStreetMap [17]. The OpenStreetMap project uses a community approach to collect geographical data of streets and buildings. Its contributors organize "mapping events" to collect the most relevant data within a predefined area in a couple of days, which has proven to be highly effective. One of the core features that makes OpenStreetMap so successful is that anybody can edit the database. Some power grid elements have already been mapped in OpenStreetMap. However, most of the required data on the sub-transmission and distribution system level is missing. Furthermore, the elements that have been mapped are often not tagged correctly, e.g., transformers are tagged as substations. For the purpose outlined above we need highly reliable data. Thus, a method for verifying incoming data is required. In OpenGridMap, we plan to address this by exploiting features of contemporary mobile devices. The collection of data using smartphones has gained a lot of momentum over the last years and is usually referred to under the umbrella of participatory sensing [21]. The main challenge that the OpenGridMap project faces is the definition of incentive schemes to kick-start and sustain the collection of relevant data by contributors. Currently, we plan to use a community approach similar to the one used by OpenStreetMap.

3   The OpenGridMap Project

One of the main challenges for Smart Grid researchers and industry experts is the access to realistic power grid data. We believe that the access to more accurate and realistic data will lead to tailored solutions that address specific issues on the current power grid. With this in mind, we launched the project OpenGridMap. Its goal is to create an open platform for inferring realistic power grids based on available and crowdsourced data. OpenGridMap has several core features: data collection, verification, inference and analysis (cf. Fig. 1). In the following, we describe these features in more detail. We plan to have several

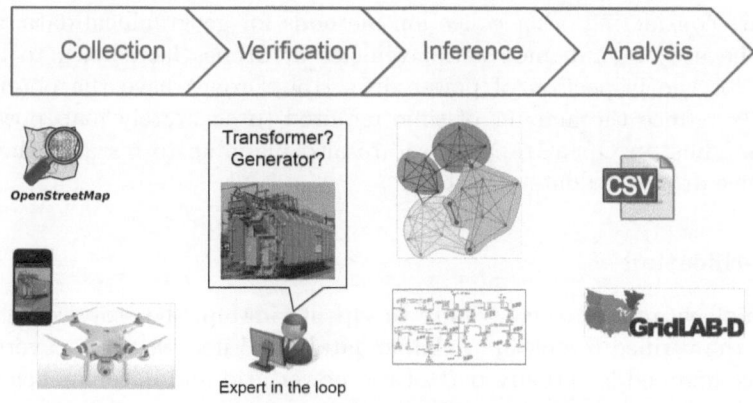

Fig. 1. OpenGridMap features

feedback loops between the core features, in particular between analysis and inference, and between analysis and collection.

3.1 Data Collection

OpenGridMap uses freely available geographical data sources, namely the ones mentioned in Subsect. 2.3. Nonetheless, as we have previously mentioned, this data is neither complete nor can its correctness be guaranteed.

Therefore, OpenGridMap collects data using a crowdsourcing approach. To facilitate this approach on the contributors' side, we offer a mobile data collection application (a mobile "app") specially tailored for the collection of power grid data. Contributors can use this app to take a picture of a potential power grid component they see and annotate it. The app adds the geographic coordinates of the picture and sends the data to the OpenGridMap's backend.

To develop a community around OpenGridMap, we plan to develop gamification schemes which provide incentives for contributors to actively participate in the project. We plan to use a combination of OpenStreetMap's community schemes and the gamification approaches used in location-based games like Geocaching [25] or Ingress [20]. Geocaching is a pastime, where a container holding a number of items is virtually hidden in a particular location for GPS users to find by means of coordinates posted on the Web. The players are rewarded with the possibility of finding an object in the container. Ingress is an augmented reality massive multi-player online role-playing game, where players are organized in two teams and battle over the control of certain key locations marked on a map that correspond with actual geographic locations. The continuous success of these two games shows that gamification is an effective method for building and sustaining a user community without having to provide monetary incentives. We therefore believe that this could be an effective approach to foster participation in OpenGridMap.

We are considering other collection methods for geographical data, as well. Lately, the use of unmanned aerial vehicles or drones has shown to be very effective for the inspection of power lines [16]. Drones have the potential to drastically reduce the amount of time required to accurately map power grid elements. Thus, in OpenGridMap we are also planning to research the use of inexpensive drones for data collection.

3.2   Verification

Since we allow anyone to contribute to OpenGridMap, the crowdsourced data needs to the verified to obtain a reliable database. First, we need to verify that the device mapped is actually part of the power grid and does not belong to a different infrastructure, such as the telecommunication network. Unfortunately, the classification of the mapped device and correct tagging is not trivial task for a common contributor. To address this issue, we envision the participation of experts in the loop, who review the submitted data and either discard incorrectly mapped elements or correct their classification. To facilitate the experts' task, OpenGridMap provides a visualization of mapped devices that includes the pictures taken by the contributors. We also plan to automate the verification process, at least partially using machine learning algorithms. For instance, these algorithms could take advantage of the fact that elements of the power grid usually carry high voltage warning signs.

3.3   Inference

Via its crowdsourcing approach, OpenGridMap has the potential to collect the visible power grid infrastructure. However, a major share of the power grid's infrastructure is not visible. Furthermore, many details cannot be determined by visual inspection only. To still be able to obtain all data required for simulation purposes, the missing data has to come from other sources or be inferred in a reliable way.

A key working area in the OpenGridMap project is topology inference, i.e., the inference of the location and technical features of distribution grid power lines. In Germany and many other countries, distribution lines are usually underground and thus cannot be visually mapped. However, based on data that we can obtain from other sources, in particular OpenStreetMap and crowdsourced data, it is possible to approximate power grid topologies based on certain background knowledge. The data collection and verification components of OpenGridMap should be able to deliver a dataset containing the location of most visible power grid elements, e.g., transformers, substations, poles, and cabinets. In addition, the system can access the geographic data provided via the OpenStreetMap component, in particular the location of streets and buildings. The inference task consists in determining the connections between the mapped power grid elements and the loads such that the overall grid topology complies with power grid design criteria. These design criteria might be different from one country to another, although several construction rules should be generalizable. In Germany, for instance, important guidelines for the design and construction of power lines are provided by the "Verband Deutscher Elektrotechniker" (VDE) [24]. To solve the resulting topology inference problem, we are currently considering three methods, including combinations of them: complex network theory, rule-based inference, and optimal network planning.

Several studies have applied complex network theory to power grid topology inference, but only, on the transmission level [8, 11, 19]. Unfortunately, the methods proposed in these papers cannot be directly applied to distribution grids, because their topological features are fundamentally different compared to transmission networks. Similar studies focussing on the distribution level are still missing.

The idea behind rule-based inference of power grid topologies is to connect given power grid elements based on common rules. Such rules can be extracted from various sources, e.g., expert documents [24], legal documents [14], economic calculations, or simply analysis of accessible power grid infrastructure plans. For instance, it is common to have power lines running along streets to enable easy access to the for maintenance and repairs.

The optimal planning approach is based on the formulation of an optimization problem that represents the design restrictions of power grids and uses standard optimization techniques to construct optimal grids given an objective, e.g., cost minimization.

To evaluate our inference methods, we plan to compare our results to actual distribution grid data. In case we cannot obtain enough of this data to perform reliable quality assessments, we plan to let experts inspect the results and judge their quality. We are also considering judging the quality of the inferred grid data based on the simulation results we are able to produce using the data.

To conduct power flow simulations, we not only need the topology of power lines, but also the characteristics of different power grid elements, e.g., transformer ratings, admittances of power lines, and the generation and load within a feeder over time. In case this data is not available, we will plan to make assumptions based on rules and expert knowledge. For instance, there are several criteria for defining the power rating of a transformer based on the number of loads connected to it. Furthermore, there exist standard procedures for choosing the type of cable to use for a particular demand level. Moreover, the design of power grids is also based on assumptions about future (maximum) demand.

3.4   Analysis

OpenGridMap will enable various types of analysis, including but not limited to power flow analysis. To facilitate power flow analysis, OpenGridMap will produce data representations of power grids that can be used directly by existing power flow tools, such as GridLab-D [10] or Matpower [30]. We plan to provide tools for categorizing inferred power grids based on a large number of metrics, e.g., size, location, population density, etc. This will, for instance, help OpenGridMap users find feeder subsets that are particularly relevant for the type of study they conduct.

4 Components

The development of OpenGridMap has just started. Nonetheless, we have already made several advances regarding collection, verification, and inference features. In the following, we present the current state of several components that our project team is working on.

4.1 Smartphone App

A first version of the OpenGridMap app that contributors can use to collect grid data is called "grid2osm GisApp" [5]. Currently, the app is only available for smartphones and tablets running Android. It can be downloaded without charge from Google's play store. The decision to implement the app in Android first instead of iOS was based its greater market share and fewer number of restrictions imposed on development.

As shown in Fig. 2a, the user interface works using gestures and currently allows the user to do the following:

- Switch to the native camera app to take a picture and record GPS coordinates of the photographed power grid element (swipe left)
- Preview the pictures taken (swipe right)
- Delete the pictures taken (swipe down)
- Store the mapped element in a submission queue (swipe up)
- Submit the picture to OpenGridMap's backend (swipe up again)

With the procedure above, users are able to map power grid elements and take multiple pictures of them. It determines the geographic location of the grid element by recording the GPS coordinates of each picture taken. Textual data is store in a sqlite database, while photos are kept on the external storage devices. In the final step of the above procedure, both textual and binary data is sent to the backend using REST.

One important challenge when using smartphones for mapping is that they offer a maximum GPS accuracy of about 10 m. Thus, our app allows for multiple pictures to be taken. We can thereby triangulate a more accurate position of the mapped element. In case this does not work, we can always rely on the expert to correct the element's position, which also becomes easier with more geolocated pictures.

A survey on the usability of the current app using the Attrakdiff questionnaire service was conducted [2]. The questionnaire features 28 pairs of opposite objectives. 20 persons anonymously participated in the survey. The results revealed that the app's quality ratings are located in the average range. It is thus clear that improvements of the app's usability are required to increase its chances of success, since usability is usually strongly related to actual usage. Currently, we are working on a second version of the app that will significantly improve usability and also include gamification elements that are supposed to provide non-monetary incentives to users.

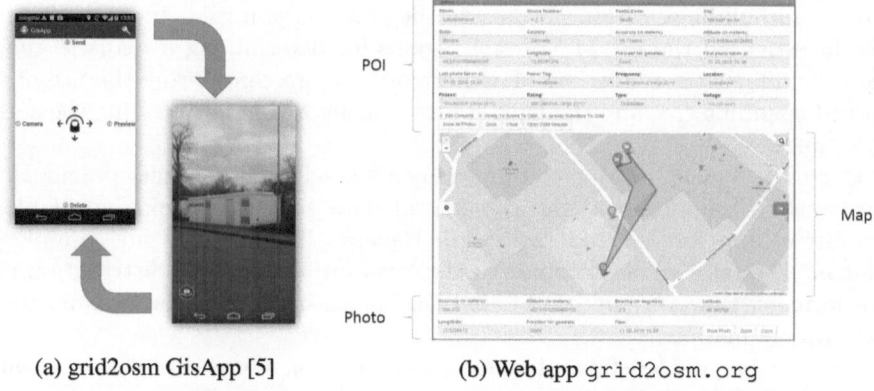

(a) grid2osm GisApp [5]   (b) Web app grid2osm.org

Fig. 2. Current OpenGridMap components

4.2   Web Application

The current OpenGridMap prototype features a web application that allows users to visualize the power grid elements currently stored. It allows for visualizing and editing device-specific data, as well as contributing it to OpenStreetMap.

Figure 2b shows how the user can access this data using the current version of the web application. Blue input fields contain location data and black input fields contain specific element data. The map based visualization pinpoints the stored GPS coordinates of the pictures taken. In case multiple pictures were taken, the application draws a polygon that visualizes the movement of the smartphone while taking pictures. Furthermore, all stored pictures can be viewed using the web application.

The primary purpose of the web application is to support experts in verifying and, if necessary, modifying the location and information fields. The status of the data set describing each grid element can be "edit complete", "ready for submission to OSM", or "submitted to OSM". The tagging options for power system elements comply with OpenStreetMap's tagging standards.

4.3   Power Grid Inference

The starting points of power grid inference are the positions of substations and distribution level transformers. Substations are the connection points between the high-voltage transmission level of the grid, whereas distribution level transformers represent the interface between the mid- and low-voltage sections of the distribution grid. We also know that most buildings are loads, i.e., they consume power at low voltage, and that most distribution grid feeders have a tree-like structure i.e., under normal operating conditions, there is usually just one path leading from each building to a substation.

Based on the data describing the geographic location and type of grid components, we have started to investigate methods for inferring tree topologies that connect all buildings within an area to the transmission grid. In particular, we have investigated the use of Voronoi regions for determining a realistic allocation of buildings to substations. Furthermore, we are considering the use of the shortest spanning tree and the Dijkstra routing algorithm for inferring the actual power lines.

Currently, OpenGridMap contains only a few self-collected data points. However, we have imported all the power grid data that is currently available in OpenStreetMap for selected regions in Bavaria. This data is not completely reliable, especially when it comes to the location and type of distribution grid transformers. However, we have found that at least the position of substations is accurately mapped.

We have begun to define inference areas based on Voronoi partitions around substations. Voronoi partitions separate an area into smaller regions based on the distance to points in a specific subset of the area. These points are also known as seeds. For each seed, the Voronoi algorithm creates a corresponding region consisting of all points closer to that seed than to any other. The resulting regions are called Voronoi cells. In Fig. 3b one can see the Voronoi cells that result when we use the mapped substations as seeds for a major German city, and the distribution transformers as seeds for a village. In both cases, one can see that the closer one gets to the center of the urban area, the smaller the cells become. This is a result of more substations or transformers situated in the center of a populated area, which in turn is (most likely) a result of larger buildings and thus a higher maximum load in the center versus the outskirts. Thus, using Voronoi regions as starting point for grid inference algorithms could be useful because it implicitly considers load distribution.

Based on an initial allocation of buildings to transformers, and transformers to substations, one can start to infer the actual power lines connecting the loads. We have started to investigate simple algorithms to perform this task. For instance, in Fig. 4a, one can see the result of a Minimum Spanning Tree (MST) algorithm applied to a fraction of the village area. Spanning trees are an interesting starting point because they are both trees and implicitly minimize the total length of the power lines. However, as one can readily observe, the results are hardly realistic, because the power lines are not aligned with the surrounding structures.

Another idea we have come up with is to take the position of streets into account. We thus implemented a grid inference method based on the Dijkstra routing algorithm for finding the shortest path that connects a transformer with buildings while following along the streets. The result can be seen in Fig. 4b. It looks much more promising, but certainly requires many adaptations to yield realistic results.

We are currently working in several directions to improve the grid inference methods. One the one hand, we are trying to combine the strengths of different algorithms, such as MST and routing algorithms. On the other hand, we want to consider more data in the inference process, e.g., available data on the size or use of buildings.

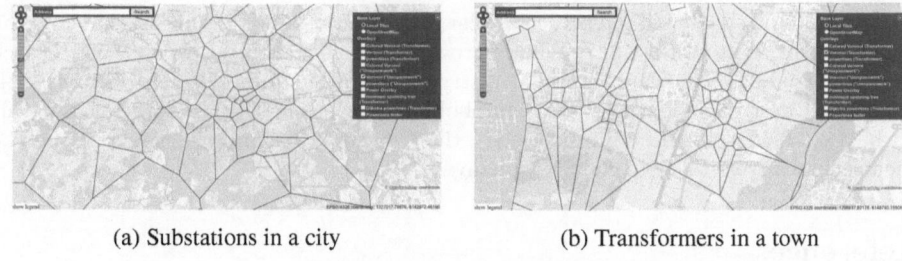

(a) Substations in a city   (b) Transformers in a town

Fig. 3. Voronoi partitions based on power grid devices location

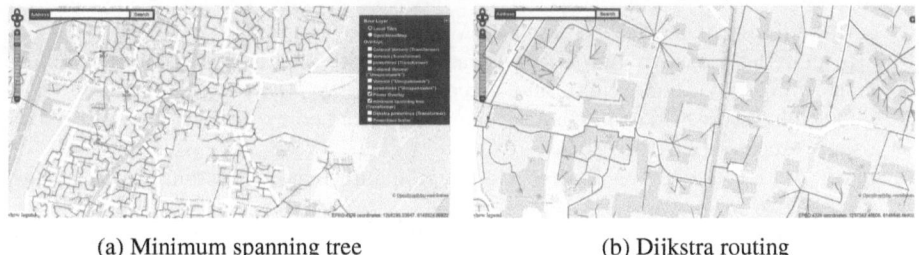

(a) Minimum spanning tree   (b) Dijkstra routing

Fig. 4. Rule-based inference of power distribution network (village)

5 Outlook

This paper presents the purpose and current status of the OpenGridMap project. The goal of OpenGridMap is to provide researchers and practitioners with detailed, realistic, and readily usable data about the existing power grid infrastructure.

We discuss the technical landscape of the project and show how Open-GridMap will advance the state-of-the-art in power grid analysis and simulation. In particular, we discuss the range of different features that OpenGridMap will provide, and which cover data collection, verification, inference, and analysis. We describe the currently available prototype, which is freely available for use. The latest developments of the project can be found at opengridmap.com. OpenGridMap is also an open-source project, i.e., its entire source code can be accessed at github.com/opengridmap.

We believe that the access to more accurate and realistic power grid data will inform more efficient decision making in the context of the energy transition. It will also enable more relevant research in the area of Smart Grids, since researchers can use the data generated by OpenGridMap to evaluate their solutions in a realistic context.

Acknowledgement. We would like to thank Klaus Schreiber, Tanuj Ghinaiya, Clotilde Guinard and Shota Bakuraze from TU München for their contributions to the project. We would also like to thank Michael Metzger from Siemens AG for his contributions as advisor to the project. Most importantly, we would like to thank all the countless contributors that have helped crowdsource geographical data. This research was supported by a German Federal Ministry of Education and Research grant (BMBF 01IS12057) and the Alexander von Humboldt Foundation. During the cause of this work, H.A. Jacobsen held affiliations with the University of Toronto, Canada, and the Technische Universität München, Germany.

References

1. ArcGIS platform Esri. http://www.esri.com/software/arcgis. Accessed 26 May 2015
2. Attrakdiff. attrakdiff.de. Accessed 26 May 2015
3. Energie-Atlas Bayern. www.energieatlas.bayern.de. Accessed 30 September 2010
4. EnergyMap. www.energymap.info. Accessed 26 May 2015
5. Google play store grid2osm GisApp. market.android.com/details?id=org.grid2osm. gisapp. Accessed 26 May 2015
6. QGIS A Free and Open Source Geographic Information System. www.qgis.org. Accessed 26 May 2015
7. Usage Limits for Google Maps API Web Services. developers.google.com/ maps/documentation. Accessed 26 May 2015
8. Albert, R., Albert, I., Nakarado, G.L.: Structural vulnerability of the North American power grid. Phys. Rev. E **69**(2), 025103 (2004)
9. Bazilian, M., Rice, A., Rotich, J., Howells, M., DeCarolis, J., Macmillan, S., Brooks, C., Bauer, F., Liebreich, M.: Open source software and crowdsourcing for energy analysis. Energy Policy **49**, 149–153 (2012)
10. Chassin, D., Schneider, K., Gerkensmeyer, C.: GridLAB-D: an open-source power systems modeling and simulation environment. In: Transmission and Distribution Conference and Exposition, T&# x00026; D. IEEE/PES, pp. 1–5. IEEE (2008)
11. Cloteaux, B.: Limits in modeling power grid topology. In: Proceedings of the 2nd IEEE Network Science Workshop, NSW 2013, April 29–May 1, 2013, Thayer Hotel, West Point, NY, USA, pp. 16–22 (2013). http://dx.doi.org/10.1109/NSW.2013. 6609189;http://dblp.uni-trier.de/rec/bib/conf/nsw/Cloteaux13
12. Deutsche Energie-Agentur: dena-Verteilnetzstudie: Ausbau- und Innovationsbedarf der Stromverteilnetze in Deutschland bis 2030. Technical report, November 2012
13. Esri: Enterprise GIS and the Smart Electric Grid. Technical report (2012)
14. Federal Minestry for Justice and Consumer Protection: Energiewirţschaftsgesetz (EnWG) (2005)
15. Gibson, R., Erle, S.: Google Maps Hacks: Tips & Tools for Geographic Searching and Remixing (Hacks)
16. Golightly, I., Jones, D.: Visual control of an unmanned aerial vehicle for power line inspection. In: Proceedings 12th International Conference on Advanced Robotics, ICAR 2005, pp. 288–295. IEEE (2005)
17. Haklay, M.M., Weber, P.: OpenStreetMap: user-generated street Maps. IEEE Pervasive Comput. **7**(4), 12–18 (2008). http://dx.doi.org/10.1109/MPRV.2008.80
18. Heipke, C.: Crowdsourcing geospatial data. ISPRS J. Photogrammetry Remote Sens. **65**(6), 550–557 (2010)
19. Hines, P., Blumsack, S., Cotilla Sanchez, E., Barrows, C.: The topological and electrical structure of power grids. In: 2010 43rd Hawaii International Conference on System Sciences (HICSS), pp. 1–10, January 2010

20. Hodson, H.: Google's ingress game is a gold mine for augmented reality. New Sci. **216**(2893), 19 (2012)
21. Kanhere, S.S.: Participatory sensing: crowdsourcing data from mobile smartphones in urban spaces. In: Hota, C., Srimani, P.K. (eds.) ICDCIT 2013. LNCS, vol. 7753, pp. 19–26. Springer, Heidelberg (2013)
22. Kersting, W.: Radial distribution test feeders. IEEE Trans. Power Syst. **6**(3), 975–985 (1991)
23. Meehan, B.: Modeling Electric Distribution with GIS. Esri Press, Redlands (2013)
24. Nagel, H., Cichowski, R.R.: Systematische Netzplanung, 2nd edn. VDE Verlag (2008)
25. O'Hara, K.: Understanding geocaching practices and motivations. In: Proceedings of the SIGCHI Conference on Human Factors in Computing Systems, CHI 2008, pp. 1177–1186. ACM, New York (2008). http://doi.acm.org/10.1145/1357054.1357239
26. Pacific Northwestern National Laboratories: Modern Grid Initiative Distribution Taxonomy Final Report. Technical report (2008)
27. Strunz, K., Fletcher, R., Campbell, R., Gao, F.: Developing benchmark models for low-voltage distribution feeders. In: Power Energy Society General Meeting, PES 2009, pp. 1–3. IEEE, July 2009
28. U.S. Department of Energy: Grid 2030: A National Vision for Electricity's Second 100 Years (2003)
29. U.S. Department of Energy: 2014 Smart Grid System Report (2014)
30. Zimmerman, R.D., Murillo-Sánchez, C.E., Thomas, R.J.: MATPOWER: Steady-state operations, planning, and analysis tools for power systems research and education. IEEE Trans. Power Syst. **26**(1), 12–19 (2011)

Towards Realistic Flow Control
in Power Grid Operation

Tamara Mchedlidze, Martin Nöllenburg, Ignaz Rutter, Dorothea Wagner,
and Franziska Wegner$^{(\boxtimes)}$

Karlsruhe Institute of Technology, 76131 Karlsruhe, Germany
{tamara.mtsentlintze,martin.noellenburg,ignaz.rutter,dorothea.wagner,
franziska.wegner}@kit.edu
http://i11www.iti.uni-karlsruhe.de

Abstract. Power flow control units (like FACTS) offer an increased
controllability and may help to tackle problems like load distribution
in future power grids. However, these control units are an expensive
investment. We showed in our previous work [10] that placing few flow
control units on buses achieves high controllability. However, current
control units are placed on transmission lines rather than buses, which
weakens the previous result. Therefore, we translate the models and
graph-theoretic explanation to control units placed on branches. Using
IEEE benchmark data, we show experimentally that few controllers on
branches still suffice to achieve minimum possible operation cost. In addition, we show that when increasing the loads, adding a small number of
control units on branches—a number comparable to the previous result—
reduces the operation costs and increases the feasibility range.

Keywords: Hybrid power flow model · FACTS · Transmission network
control · Graph theory

1   Introduction

An increasing number of renewable energy producers is added to the power grid
to increase sustainability. Renewable energy producers are often independent
power producers (IPPs), which represent an external power supply in contrast
to the classical power supply. IPPs cause energy flow patterns that differ from
what the classical power grid was designed for, since they can be placed in the
low-, medium- and even high-voltage grid. There are two strategies for extending
current power grids to ensure reliable and cost-efficient energy supply also in the
presence of these changes.

(S1) Extending the grid with additional transmission lines.
(S2) Installing control units like flexible AC transmission systems (FACTS) [8]
     to enhance grid utilization.

This work was funded (in part) by the Helmholtz Program Storage and Cross-linked
Infrastructures, Topic 6 Superconductivity, Networks and System Integration.

© Springer International Publishing Switzerland 2015
S. Gottwalt et al. (Eds.): EI 2015, LNCS 9424, pp. 192–199, 2015.
DOI: 10.1007/978-3-319-25876-8_16

In this paper, we consider the latter strategy. We study the positive effects of placing control units on selected branches for the operation cost and operability. It has been shown in [10] that if a flow control unit is placed on every branch, the operation cost decreases and represents the lower bound for the operation cost. In this work, we analyze the minimum number of flow control units placed on branches (rather than on buses) necessary to achieve the lower bound for the operation cost. We analyze the positive effect of placing control units on branches in terms of grid operability in the presence of increasing grid load. Finally, we compare our results to the model given in [10].

This work is an improvement of our previous model [10], where control units are placed on buses having an effect of controlling all their incident edges. Here, we make this model more realistic by considering control units being placed on branches instead of buses. We assume that a flow control unit is an *ideal FACTS* [6] controlling the power flow on its branch without any restrictions.

Using the IEEE power system test cases[1] and a Python implementation of MATPOWER[2] called PYPOWER[3], we performed simulation experiments related to two key questions.

(Q1) How many and on which branches flow control units need to be placed to obtain the lower bound for the operation cost?
(Q2) We consider the state of the grid, where the branch limits are approached and ask whether a limited number of flow control units can decrease the operation costs and extend the grid operability?

In Sect. 2, we give an overview of the previous work. Our new model is introduced in Sect. 3. In Sect. 4, we analyze the Question Q1 and explain the experimental results by relating them to the theoretical facts proven in [10]. In Sect. 5, we explore Question Q2 and analyze the difference between the results of the current and the previous model [10]. We conclude and give an outlook in Sect. 6.

2   Related Work

The problem of generating the required amount of power while obtaining minimum operation cost and meeting some restrictions is called *Economic Load Dispatch Problem* (EDP). For the classical EDP a non-linear integer program needs to be solved, which is NP-hard [7]. To cope with the EDP without FACTS, the optimal power flow (OPF) method—a numerical method—was introduced by Carpentier [1]. The development of OPF with its fundamental refinements is summarized by Frank et al. [3,4].

Researchers study the advantageous effect of positioning control units like FACTS and approach Questions Q1–2 in different ways. Gerbex et al. [5] and

---

[1] data sources http://www.pserc.cornell.edu/matpower/ and http://www.ee.washing ton.edu/research/pstca/.
[2] http://www.pserc.cornell.edu//matpower/.
[3] https://pypi.python.org/pypi/PYPOWER/4.0.0.

Ongsakul and Jirapong [13] use a genetic algorithm and evolutionary programming, respectively, to optimize the positioning of FACTS by meeting different special constraints. Lima et al. [11] propose a mixed-integer linear program to optimally increase the loadability. Similar to our approach, they assume "ideal" FACTS that can arbitrarily control all transmission line parameters. Differently to our approach the authors neglect generation costs and line losses. All the aforementioned approaches use DC model as an approximation of AC model.

AC models are more realistic than their DC approximations. There exist different AC models [12] using cartesian/polar coordinates or trigonometric approximations. These models can be categorized as follows:

- AC models with sinusoidal loads (non-convex and non-linear formulation),
- AC quadratic approximations (non-convex and quadratic formulation),
- AC piece-wise-linearization (non-convex and integer linear programming formulation),
- AC linearization (convex and linear formulation).

Farivar and Low [2] present an approach for the last model. They evaluate an exact OPF model by convexifying and relaxing the AC-model. In this context, they place phase shifters to exploit structural characteristics that are similar to our approach. This is necessary to get an valid solution in the original AC-model. Zimmerman et al. [14] study the positioning of FACTS on critical lines to improve the voltage stability of power grids. Ippolito et al. [9] investigate the number and the location of FACTS to maximize the system capability.

3   Model

We use a DC power grid model and map a power grid with buses and branches, called in the following *transmission lines* or simply *lines*, to a graph $G = (V, E)$ with a set of vertices V and a set of edges $E \subseteq \binom{V}{2}$, respectively. The set of *flow control lines* (FCLs), denoted by $F_E \subseteq E$, are lines with ideal FACTS controlling flow on them. Analogously, we define the set of *flow control buses* (FCBs) and denote them by $F_V \subseteq V$. The FCBs control the flow on all incident edges. We denote by G_{F_E} the subgraph of G that contains all edges of F_E and their endvertices. Similarly, we denote by G_{F_V} the subgraph of G that contains the vertices of F_V and all incident to them edges. A *flow* f in G is a function $f \colon V \times V \to \mathbb{R}$ satisfying the flow conservation (Kirchhoff's current law, see Eq. 1b) constrained by line limits $c(u, v)$ (see Eq. 1d) and fulfilling consumer/generator constraints (see Eqs. 1e and 1f). Here, $V_G \subseteq V$ is the set of generators, $x_v \in \mathbb{R}^+$ is the maximum supply of generator $v \in V_G$, $V_C \subseteq V$ is the set of consumers and $d_u \in \mathbb{R}_0^+$ is the demand of $u \in V_C$. A function f is called *electrically feasible* if there exists a voltage angle assignment $\Theta \colon V \to \mathbb{R}$ such that Eq. 1c holds, where $B(u, v)$ is the *susceptance* of the edge (u, v).

In the standard *flow model* the flow on all edges can be manipulated, i.e., it is like having a FACTS on every edge in a power grid. In case of FCBs this means $F_V = V$ and in case of FCLs this means $F_E = E$. The standard flow model asks to find a flow, i.e., it neglects Eq. 1c. In the *electrical flow model* [15] the flow on edges cannot be controlled, which translates to $F_V = \emptyset$ in case of FCBs and $F_E = \emptyset$ in case of FCLs. The *electrical flow model* requires to find an electrically feasible flow, i.e., Eq. 1c satisfied for all edges. In the *hybrid model* introduced in [10], the flow model and the electrical flow model are combined and it is required to find a flow on G_{F_V} such that Eq. 1c holds only for edges whose both endpoints are not in F_V, i.e., the flow must be electrically feasible only on the subgraph induced by $V \setminus F_V$.

Since an *ideal FACTS* [6] is technically realized on transmission lines, it is more realistic to consider FCLs instead of FCBs. Thus, we translate the three described models designed for FCBs to models on FCLs by simply replacing F_V by F_E and G_{F_V} by G_{F_E}.

Recall that the EDP is the problem of generating the required amount of power while obtaining minimum operation cost and meeting the constraints in Eqs. 1b–1f. The objective function $\gamma_\lambda(f)$ describing the operation cost is a weighted function of generator costs $\gamma_G(f)$ and transmission line losses $\gamma_L(f)$, where λ is the weight factor (see Eq. 1a). The overall optimization problem is as follows:

$$\text{minimize } \lambda \cdot \gamma_G(f) + (1 - \lambda) \cdot \gamma_L(f) \tag{1a}$$

subject to

$$\sum_{\{u,v\} \in E} f(u, v) = 0 \qquad v \in V \setminus (V_G \cup V_C) \tag{1b}$$

$$f(u, v) = B(u, v)(\Theta(u) - \Theta(v)) \qquad \forall \{u, v\} \in E \setminus F_E \text{ for FCLs}$$
$$\forall \{u, v\} \in E \text{ s.t. } u, v \notin F_V \text{ for FCBs} \tag{1c}$$

$$-c(u, v) \leq f(u, v) \leq c(u, v) \qquad \{u, v\} \in E \tag{1d}$$

$$\sum_{\{u,v\} \in E} f(u, v) = -d_v \qquad v \in V_C \tag{1e}$$

$$0 \leq \sum_{\{u,v\} \in E} f(u, v) \leq x_v \qquad v \in V_G \tag{1f}$$

4   Evaluation of Placing Flow Control Branches

In this section, we transfer our previous theoretical results [10] to FCLs. Thereby, we answer Question Q1: How many FCLs are necessary to achieve the lower bound for the operation cost, which happens in case each line is a FCL. We call this operation cost a *full control cost*.

In Fig. 1 the graph of the IEEE `case14` with the different subgrids is shown for the placement of FCLs, that induces the operation cost equal to the full control cost. We observe that the subgrid $G - F_E$ (graph G where the edges of F_E are removed) forms a *cactus* (graph where each edge lies in at most one cycle). In other examples, we observed that $G - F_E$ can be even simpler, forming a *forest* (a graph without any cycle). If $G - F_E$ is a cactus (resp. forest) and F_E is the smallest such set, set F_E is called *minimum cactus* (resp. *forest*) *feedback set*.

In our experiments, summarized in Fig. 2, we compared the number of FCLs necessary for achieving the full control cost to the size of minimum forest and

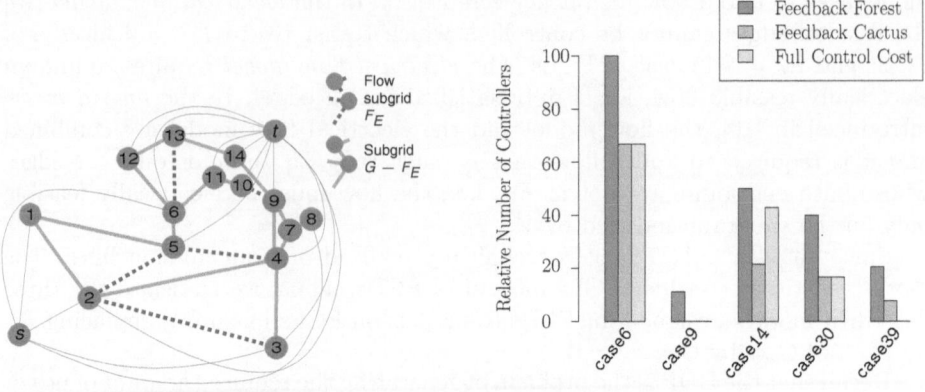

Fig. 1. IEEE benchmark `case14` including the minimum numbers of controllers for $\lambda = 0.5$, where the bold normal lines represent $G - F_E$ and the bold dashed lines represent FCLs F_E.

Fig. 2. Comparing of the minimum feedback set sizes for forests and cacti with the necessary number of FCLs for full control. Cases 9 and 39 need zero FCLs, which is equivalent to [10].

cacti feedback sets. In `case6`-`case30` the number of edges for the full control cost is between the minimum size of a forest feedback set and a cactus feedback set. In addition, `case6`, `case9` and `case30` achieve full control cost with FCL size equal to the size of a feedback cactus set. For `case39`, full control is achieved with fewer FCLs than the feedback cactus set size. Unfortunately computing the optimal number of FCLs for the larger IEEE test cases is prohibitively expensive with our current integer linear programming formulation.

The following two theorems provide theoretical evidence for our empirical observations. They explain why the number of FCLs to achieve full control cost and the size of minimum cactus/forest feedback set are related. This relation and the fact that power grids are not dense networks, i.e., their feedback forest set is not large, suggests that the relatively small number of FCLs are enough to achieve the full control cost. Farivar and Low [2] give similar structural results on spanning trees, but using a different model.

Theorem 1. *Let $G - F_E$ be a forest. Then every flow f is electrically feasible on G_{F_E}.*

Theorem 2. *Let G_{F_E} be a power grid with FCLs at the edges in F_E such that $G - F_E$ is a cactus and every edge of $G - F_E$ that lies on a cycle has infinite line limits (or suitably bounded, see [10]). For any flow f there exists a flow f' with identical cost that is electrically feasible for G_{F_E}.*

The proofs for these theorems can be directly derived from [10] by replacing the set of flow control buses F_V with the set of flow control lines F_E.

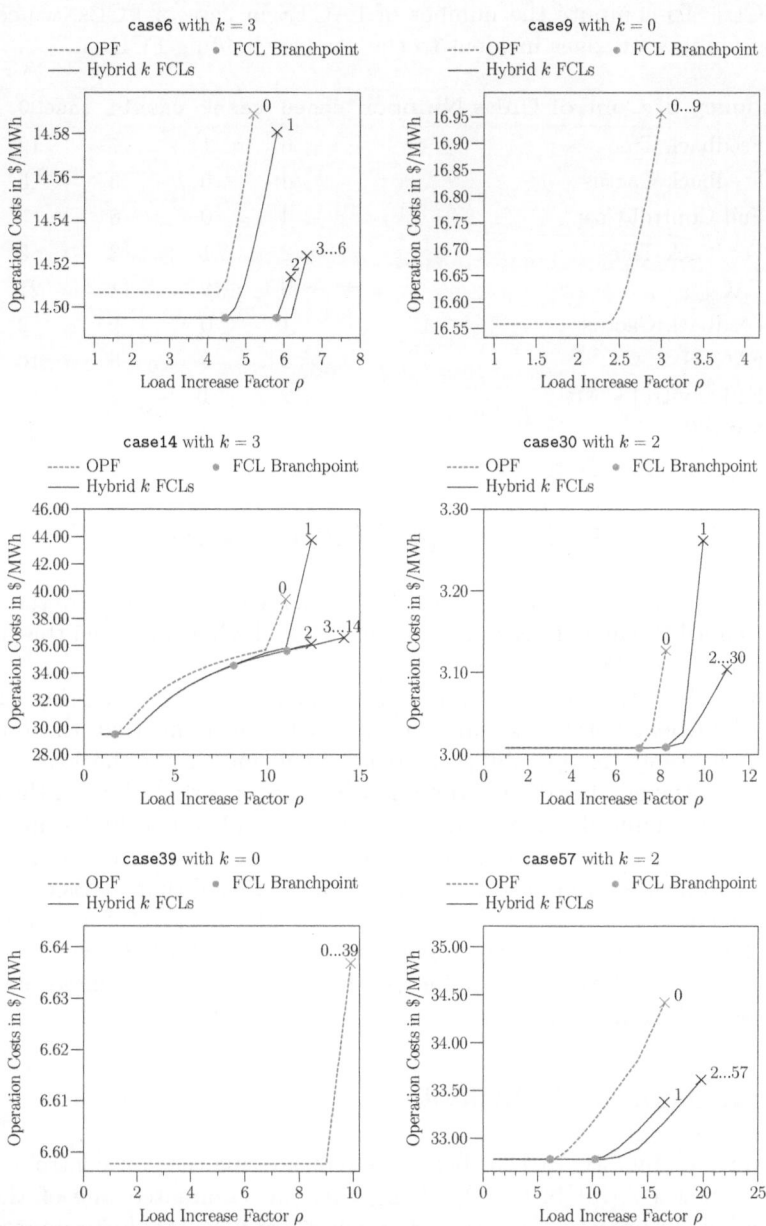

Fig. 3. Overview of operation costs for `case6` to `case57` for OPF (0 FCLs) and the hybrid model with respect to load increase factor ρ, where k is the upper bound for FCLs. The numbers on the curves represent the number of FCLs for that specific curve. Cases 9 and 39 need zero FCLs, which is equivalent to [10].

Table 1. Comparison of the previous model using FCBs and the current model using FCLs. To compute the number of FACTS in case of FCBs, we compute the total number of edges incident to the vertices holding FCBs.

Dependency of Control Units Number		case6	case9	case14	case30	case39
FCLs	Feedback Tree	6	1	7	12	8
	Feedback Cactus	4	0	3	5	3
	Full Control Cost	4	0	6	5	0
FCBs	Feedback Tree	2	1	3	5	4
	FACTS	9	2	11	21	15
	Feedback Cactus	1	0	2	2	2
	FACTS	5	0	8	10	5
	Full Control Costs	2	1	2	5	4
	FACTS	9	3	8	23	15

5   Effect of FCLs in Comparision to FCBs

In this section we evaluate Question Q2. For this reason, we increase the load by a factor ρ until the model becomes infeasible. For the hybrid model this happens when adding more FCLs does not extend the operability.

Figure 3 show the experimental results for the IEEE power grids case6 to case57. The behavior is the same as for FCBs meaning that the operation cost and the range of operability increase when increasing the load factor ρ. Interestingly, the number of FCLs does not increase substantially. For the case14 there is a maximum of three FCLs necessary instead of two FCBs and for the case57 the number of maximum FCLs remains the same as for FCBs. Similar behavior can be observed for the remaining cases. Recall that FCBs control flow on all incident edges, which can be realized by placing FACTS on all of these edges. Thus, in case of FCBs the number of necessary FACTS actually depends on the degree of the vertices holding the FCBs and results in large number of FACTS as indicated in Table 1.

6   Conclusion and Outlook

In this work, we investigated the benefits of considering the flow control branches instead of flow control buses. By doing this, we eliminated one of the main drawbacks of our previous work [10]. However, there is still room for improvement of the model. In the future, we will extend our work from DC simplification to a more realistic AC power grid model. In problems like transmission network expansion planning (TNEP), the solution of DC models may substantially differ from an AC model and is often not feasible. For this reason, the work of Farivar and Low [2] introducing an AC model simplification is a good point to start. An alternative research direction is to start with a DC linear model without losses but include reactive power. It would also be interesting to consider different AC models, e.g. with cartesian/polar coordinates or trigonometric approximation. In future work, we plan to generalize our results to these more realistic models and to evaluate which model is best for producing realistic results.

References

1. Carpentier, J.: Contribution to the economic dispatch problem. Bull. Sac. France Elect. **8**, 431–437 (1962)
2. Farivar, M., Low, S.: Branch flow model: relaxations and convexification - part II. IEEE Trans. Power Syst. **28**(3), 2565–2572 (2013)
3. Frank, S., Steponavice, I., Rebennack, S.: Optimal power flow: a bibliographic survey I. Energy Syst. **3**(3), 221–258 (2012)
4. Frank, S., Steponavice, I., Rebennack, S.: Optimal power flow: a bibliographic survey II. Energy Syst. **3**(3), 259–289 (2012)
5. Gerbex, S., Cherkaoui, R., Germond, A.: Optimal location of multi-type FACTS devices in a power system by means of genetic algorithms. IEEE Trans. Power Syst. **16**(3), 537–544 (2001)
6. Griffin, J., Atanackovic, D., Galiana, F.D.: A study of the impact of flexible AC transmission system devices on the economic-secure operation of power systems. In: 12th Power System Computer Conference, pp. 1077–1082 (1996)
7. Hemmecke, R., Kppe, M., Lee, J., Weismantel, R.: Nonlinear integer programming. In: Jünger, M., Liebling, T.M., Naddef, D., Nemhauser, G.L., Pulleyblank, W.R., Reinelt, G., Rinaldi, G., Wolsey, L.A. (eds.) 50 Years of Integer Programming 1958–2008, pp. 561–618. Springer, Berlin (2010)
8. Hingorani, N.: Flexible AC transmission. IEEE Spectr. **30**(4), 40–45 (1993)
9. Ippolito, L., Siano, P.: Selection of optimal number and location of thyristor-controlled phase shifters using genetic based algorithms. IEE Proc. Gener. Transm. Distrib. **151**(5), 630–637 (2004)
10. Leibfried, T., Mchedlidze, T., Meyer-Hübner, N., Nöllenburg, M., Rutter, I., Sanders, P., Wagner, D., Wegner, F.: Operating power grids with few flow control buses. In: Proceedings of the ACM Sixth International Conference on Future Energy Systems, e-Energy 2015, pp. 289–294. ACM, New York (2015). http://arxiv.org/abs/1505.05747
11. Lima, F., Galiana, F., Kockar, I., Munoz, J.: Phase shifter placement in large-scale systems via mixed integer linear programming. IEEE Trans. Power Syst. **18**(3), 1029–1034 (2003)
12. Momoh, J., El-Hawary, M., Adapa, R.: A review of selected optimal power flow literature to 1993. II: Newton, linear programming and interior point methods. IEEE Trans. Power Syst. **14**(1), 105–111 (1999)
13. Ongsakul, W., Jirapong, P.: Optimal allocation of FACTS devices to enhance total transfer capability using evolutionary programming. In: IEEE International Symposium on Circuits and Systems (ISCAS 2005), vol. 5, pp. 4175–4178
14. Sharma, N., Ghosh, A., Varma, R.: A novel placement strategy for FACTS controllers. IEEE Trans. Power Delivery **18**(3), 982–987 (2003)
15. Zimmerman, R.D., Murillo-Sanchez, C.E., Thomas, R.J.: MATPOWER: Steady-state operations, planning, and analysis tools for power systems research and education. IEEE Trans. Power Syst. **26**(1), 12–19 (2011)

Configuration of Hydro Power Plant Mathematical Models

Michael Barry[1], Moritz Schillinger[2], Hannes Weigt[2], and René Schumann[1(✉)]

[1] Smart Infrastructure Laboratory, HES-SO Valais/Wallis, Rue de Technopôle 3,
3960 Sierre, Switzerland
{michael.barry,rene.schumann}@hevs.ch
[2] Forschungsstelle Nachhaltige Energie- und Wasserversorgung, University Basel,
Peter Merian-Weg 6, 4002 Basel, Switzerland
{moritz.schillinger,hannes.weigt}@unibas.ch

Abstract. The ongoing energy transition towards large shares of renewable generation poses challenges for hydro power producers. We revisit the problem of optimising the operation of hydro power plants using mathematical modelling, but utilising computer science concepts in the design of the models and configuration of these models. We use a modular design allowing us to activate features, such as what markets or which technical aspects to consider, by activating or deactivating a specific module. In this paper we give an example of how our method can be used to configure which markets a model should operate on. Furthermore, we use a configuration process based on the SPEA2 evolutionary algorithm to explore the relationship between the scale of the model and the time required to solve it. Such methods assist in identifying configurations that are the best fit in terms of runtime, realism and accuracy.

1 Introduction

Due to the energy transition throughout Europe there is an ongoing shift from using conventional energy sources such as fossil or nuclear to renewable sources such as photo-voltaic or wind power. However, such power sources are intermittent due to weather conditions and therefore create larger fluctuations in the energy system. These fluctuations can also be seen in the energy prices with an additional significant decline of the average price levels and a flattening of daily price peaks due to solar injections.

Hydro power (HP) is a mature technology with over a century of utilization history and is therefore well understood, including its optimisation. Highly accurate mathematical models have already existed for some time. However, recently challenges emerged from the energy market, listed below, contributing to complicating the problem and therefore forcing us to revise our methods.

- *Less Time:* Due to the less predicable power production due to higher use of weather dependant renewable energies, and the resulting fluctuations in the market prices, the operation plans needs to be adapted quickly. As planning relevant information becomes only available short term, the method for

© Springer International Publishing Switzerland 2015
S. Gottwalt et al. (Eds.): EI 2015, LNCS 9424, pp. 200–207, 2015.
DOI: 10.1007/978-3-319-25876-8_17

operation planning has to have a low runtime. Also, the intraday market can be lucrative for HP but only allows for a short time frame for planning.

- *Larger Problem:* As the intra-day market works on 15 min time slots, with a possible reduction to 10 min, it is a more complex problem and requires more data than with a more coarse grain time discretisation.
- *Smaller Margin or Error:* Like other conventional power plants HP plants also struggle with the currently low energy prices, in particular as they have to pay fees to use the water and are exempted from most green energy benefits. This limits the profitability of a HP plant and therefore places a bigger strain on optimisation models to maximise the profit.
- *Model Maintenance:* Mathematical models exist, but like most mathematical models, they are difficult to modify to adapt to a changed environment.

One solution might seem to shift from mathematical modelling to more heuristic orientated methods. But in this field mathematical modelling is well established, and existing models should stay in use, when possible. In addition, most market models use mathematical modelling and must be compatible.

However, there are many aspects in mathematical modelling that can benefit from fundamental concepts in computer science. This includes design, development and deployment aspects. In this paper we focus on the configuration of the mathematical models. Current models are problem specific and are hard to solve. However, in some instances it can be beneficial to use a lighter model with a lower runtime, so that more up-to-date data can be used. We will provide here an approach towards a more flexible model that can adjust its scale and is highly flexible, allowing for easier modification and maintenance.

Our idea bases on the assumption that a mathematical model is based on different modules describing aspects of the problem. For each aspect alternative modules can exist. By combining different modules a complete model can be created. The combination of modules is the *configuration* of the model. Based on the idea that mathematical models are wrong, but some are useful [3], we are going to investigate the trade-off between usefulness of a model and its resulting computational complexity. Therefore, we will investigating the search space of different configurations of these models in terms of scale and runtime. To investigate we propose an easy to implement method to show the trade-off between the scale of the model and its runtime by investigating the Pareto front. The Pareto front is expected to be interesting in a research and practical sense.

The optimal configurations and the pattern in which they can be observed may be of research interest, as they may point to an easy-hard-easy pattern [6]. The aim is to identify patterns in the optimal configurations within the Pareto front for different case studies (different hydro plants) to define general guidelines that can assist in designing models in the future. This is a research in progress paper. First we discuss some related literature. In Sect. 3 the problem definition is presented, followed by our approach in Sect. 4. We then provide first results (Sect. 5), and finally outline our future work and expectation.

2   Background

Using mathematical models for the planning of the ideal operation of a hydro plant is a well known and tested method. This is the case for both single site models [5,7] and cascading (multi-site) models [1,4,9]. Even large models, such as the HP plant system in Brazil consisting of 150 hydro plants, can be modelled using this method [8]. However, there are, to our knowledge, no other works that aims to use a modular design to develop these models and utilise this design as part of a configuration process to ascertain ideal configurations. As for the configuration of the model, it is a question of how simple the model should be. Operations research is very familiar to the concept that *Essentially, all models are wrong, but some are useful.* [3] as well as with the somewhat opposing principle of Occam's razor [2], stating that with *competing hypotheses that predict equally well, the one with the fewest assumptions should be selected.* For the field of operations research, it is often difficult to justify a model to be simple but accurate enough expect through practical testing. In this paper, we attempt to create a more systematic approach. In addition, through identifying the Pareto front, we hope to find an easy-hard-easy pattern [6].

The configuration is multi-objective as we aim to minimise the required runtime to solve the model and reduce the scale of the model. Multi-objective problems usually consider the Pareto front as a solution, which are proved to be NP-hard to compute [10]. Therefore heuristics are commonly used, of which evolutionary methods have proven to be effective off-the shelf algorithms [10].

3   Problem Definition

3.1   Optimisation Problem

The optimisation problem of a HP plant can be defined in relatively simple terms and is similar to a mathematical representation of a battery. However, we must consider that we have the possibility of trading on several markets. In this paper we consider the optimisation problem in its basic form for simplicity. However, we consider the problem to be scalable, as many technical, environmental and market constraints can be added. The basic form shown below:

$$\max. \sum_{i,m} P_{i,m} Q_{i,m} \tag{1}$$

$$Q_{i,m} = R_{i,m} \alpha \tag{2}$$

$$S_i = S_{i-1} + I_i - \sum_m R_{i,m} \tag{3}$$

$$S_i \leq S_{\max} \tag{4}$$

$$S_i \geq S_{\min} \tag{5}$$

$$\sum_m Q_{i,m} \leq Q_{max} \qquad (6)$$

Where $P_{i,m}$ is the price at time interval i for market m, $Q_{i,m}$ is the produced energy for time interval i and market m, $R_{i,m}$ is the water released from the reservoir at time interval i for market m, α is the efficiency of the turbine (the amount of energy produced per water used), S_i is the storage level at time interval i, I_i is the inflow of water into the reservoir at time interval i, $S_{\max}$ is the maximum storage level of the reservoir, $S_{\min}$ is the minimum storage level of the reservoir and Q_{max} is the maximum production level.

3.2   Configuration Problem

Below we describe the configuration of the model used to solve the optimisation problem defined above. Each configuration is defined by which markets are used. Using all markets increases the scale of the model and therefore directly affects the time required to solve it. The configuration problem can be summarised as finding the Pareto front of this trade off. The problem is defined formally below.

Given a configuration x in the set of all possible configurations X and functions $f_i(x)$ for each objective i, x is said to be Pareto optimal if no other member $x*$ of X dominates x. $x*$ dominates x if the following is true:

$$f_i(x) \leq f_i(x^*) \text{ for all } i \text{ where } f_i(x) < f_i(x^*) \text{ for at least one objective } i \qquad (7)$$

For the configuration of the HP model, there are two objective $f_r(x)$ and $f_s(x)$ where $f_r(x)$ represents the runtime of configuration x and $f_s(x)$ represents the scale of the model in configuration x.

4   Method

This section describes our own approach to address the previously described problems, utilising the General mathematical Modelling System (GAMS), the IBM CPLEX solver and the SPEA2 evolutionary algorithm [11].

4.1   Mathematical Model Design

As previously described, we wish to create scalable models that can be easily configured. An object orientated inspired design was used to map functions to features. Each feature is represented by a module, containing all associated mathematical functions. Each module is then stored in a separate file. A main file containing a list of import statements can then be adjusted to simply exclude a file and therefore it's feature. For example, each function for the intraday market is contained in a module and therefore it is possible to switch of trading on the intraday market by excluding the intraday market module. As the model grows, new modules are added that either replace modules or compliments them. Additional environmental constraint can be written in new modules and added, while

Data: N: population size, $\overline{N}$: elite population size, T Max number of generations

Result: $\overline{P}_t$ the non-dominated set

Initialise population P_0 with randomly generated individuals and empty set $\overline{P}_0$

Set t=0

while $t < T$ **do**

  for all x in P_0 and $\overline{P}_0$ **do**

    $f(x) = N - N_d$ where $f(x)$ is the fitness function and N_d is the number by which the individual in dominated by

  end for

  Move non-dominated individuals from P_t and $\overline{P}_t$ to $\overline{P}_{t+1}$.

  if size of $\overline{P}_{t+1} > \overline{N}$ **then**

    use clustering method to reduce the size of $\overline{P}_{t+1}$

  end if

  for all x in P_t and $\overline{P}_t$ **do**

    move P_t to P_{t+1} based on roulette wheel selection

  end for

  while Size of $P_{t+1} < N$ **do**

    breed x^* from $\overline{P}_t$ and P_t

    add x^* to P_{t+1}

  end while

  t = t + 1

end while

return $\overline{P}_T$

Algorithm 1. SPEA2 algorithm

a new turbine for a new case study can be written and then added in a simulation for the new case study. The entire model is implemented using GAMS and once configured, CPLEX is used to solve it. This modular design has many benefits similar to object orientated design, including abstraction, mapping modules or real life objects, better maintainability and deployment.

4.2 Configuration

Each set of modules used is considered to be a configuration. A list of modules that are required for the basic model to run for a particular case study is used to ensure those models are always switched on and the remaining modules are considered by the configuration process. The configuration process is based on the SPEA2 algorithm. It uses Pareto domination in it's fitness function, a separate population of non-dominated individuals to implement elitism and a clustering algorithm to stop convergence. Details are given in the listing of Algorithm 1.

As mentioned, we use the runtime and scale of the model as objectives. As the runtime is dependent on the hardware used and other software running, we use the CPLEX ticks as a platform independent measure of runtime. We use the number of modules as a preliminary measure of the scale of the model. Although this is a relative simple method it has shown to be fairly dependable. In the future, we plan to exchange this measure with more sophisticated ones.

We have chosen an evolutionary algorithm for the following reasons. First, a population based approach is well suited for finding the Pareto front, as an entire set of Pareto optimal solutions are contained in the elite population, reducing the number of requiring reruns. Second, an iterative population based approach effectively investigating an unknown search space, as it provides the stability of the scale measurement by identifying if related individuals also have similar runtimes. In addition, evolutionary algorithms are highly configurable, allowing us to update separate components, such as the initialisation of the population, to achieve better performance.

5   Initial Results

Results of solving the mathematical model are shown in Fig. 1(a). We added markets in order, therefore a model with 4 markets has markets 1–4 activated. Each market has a separate implementation including the day-ahead, intra-day, primary and secondary reserve, positive and negative tertiary reserve market (market 1–6). These markets modules are test implementations and contain fictive data and we do not consider the dependency between markets for simplicity.

In general, it shows how increasing the number of markets that the model is able to trade on also increases the revenue, showing that there is a clear profit gain in being able to trade on all markets. However, market 2 appears to never be favourable which is why trading in one or two markets in Fig. 1(a) shows no difference. Trading in all 6 markets or all expect market 2 also have the same results (not shown in Fig. 1(a)). Thus, it may be of use to exclude market 2 from the model. However, as market data is faked, it just showcases the potential of our approach.

In this section we present our initial results. These results are only viable as a proof of concept, but are not yet complete, mostly due to the fact that the model is still at an early stage and therefore is simply not large enough yet to create a large enough search space for the evolutionary algorithm. For this section we use a small scale model as a proof of concept, with simple implementations of the technical aspects and several markets to choose from. The model is scalable only through the choice of which and how many markets to trade on. The evolutionary algorithm may seem obsolete in these initial results due to the small scale, especially as the clustering algorithm is never activated due to the limited number of possible Pareto optimal solutions. Figure 1(b) shows the final results, outlining the Pareto front. Figure 1(b) demonstrates a clear relationship between the scale measurement and runtime, which demonstrates the feasibility of using the number of modules as a measurement of scale. It shows that configurations with similar scale also have a similar runtime and that there is a proportional relationship between the runtime and the scale. However, more details about how stable and what type of relationship exists will only be obtained once a large model is implemented.

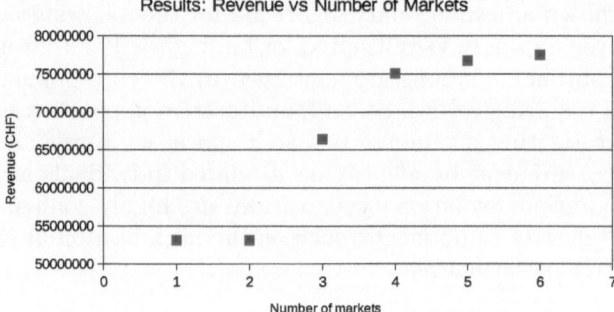

(a) Revenue for different market configurations

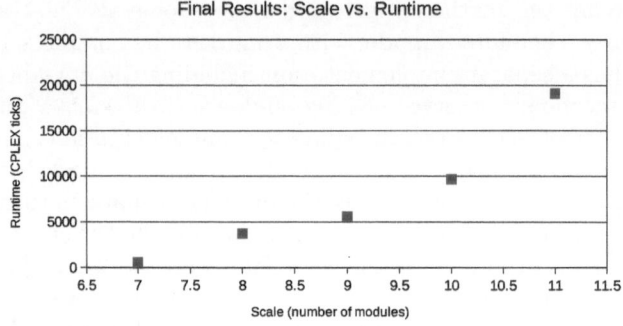

(b) Pareto front of time and scale for different configurations

Fig. 1. Initial results of the configuration process, showing the pareto front

6 Future Work and Expectations

The research project is still in an early stage. In the next steps we will focus on expanding the mathematical model, greatly increasing the search space of the configuration problem. As we are going to apply our approach in real-world case studies, competing with industry standards, we have to implement more technical and environmental constraints and more sophisticated market models. In total the project will contain two case studies, first a single HP plant, and later cascading HP plants, for which operation planning has to be done in one planning process, due to their physical dependencies.

Beside increasing the complexity of the model, we also need to revise our measurements for the size of the problem. The currently used number of modules is just an intermediate step, and more sophisticated models needs to be defined.

Once the model is of a larger scale, we will improve the configuration. There are large improvements that can be done for the evolutionary algorithm, especially once we understand the search space better. Knowledge of previous runs can be used in the initialisation of the population, reducing the time to find the Pareto front. To assist in the population spreading along the Pareto front a modified mutation operator can be used to favour mutation to a larger or smaller scale model. Additionally, the fitness evaluating requires to solve the model and therefore is time consuming. To speed-up evaluation, a hash map can be used to look-up solutions instead of recomputing them.

We also need to analyse in more depth the search space, especially whether the stability we observed above also exists in larger models, and whether the simple relationship between scale and runtime remains or if fluctuations or even easy-hard-easy curves can be observed. This insights can be used to identify areas that promise to have a low runtime despite a relatively large model.

Acknowledgements. This work has been done in the context of the SNSF funded project *Hydro Power Operation and Economic Performance in a Changing Market Environment*. The project is part of the National Research Programme *Energy Transition* (NRP70).

References

1. Alfieri, L., Perona, P., Burlando, P.: Optimal water allocation for an alpine hydropower system under changing scenarios. Water Resour. Manage. **20**(5), 761–778 (2006)
2. Blumer, A., Ehrenfeucht, A., Haussler, D., Warmuth, M.K.: Occam's razor. Inf. Process. Lett. **24**(6), 377–380 (1987)
3. Box, G.E., Draper, N.R.: Empirical Model-Building and Response Surfaces. Wiley, New York (1987)
4. Guo, S., Chen, J., Li, Y., Liu, P., Li, T.: Joint operation of the multi-reservoir system of the three gorges and the Qingjiang cascade reservoirs. Energies **4**(7), 1036–1050 (2011)
5. Duque, Á.J., Castronuovo, E.D., Sánchez, I., Usaola, J.: Optimal operation of a pumped-storage hydro plant that compensates the imbalances of a wind power producer. Electr. Power Syst. Res. **81**(9), 1767–1777 (2011)
6. Mammen, D.L., Hogg, T.: A new look at the easy-hard-easy pattern of combinatorial search difficulty. JAIR **7**, 47–66 (1997)
7. Pérez-Díaz, J.I., Wilhelmi, J.R., Arévalo, L.A.: Optimal short-term operation schedule of a hydropower plant in a competitive electricity market. Energy Convers. Manage. **51**(12), 2955–2966 (2010)
8. Zambelli, M., Huamani, I., Kadowaki, M., Soares, S., Ohishi, T.: Hydropower Scheduling in Large Scale Power Systems. INTECH Open Access Publisher, Rijeka (2012)
9. Zhang, X.M., Wang, L.P., Li, J.W., Zhang, Y.K.: Self-optimization simulation model of short-term cascaded hydroelectric system dispatching based on the daily load curve. Water Resour. Manage. **27**(15), 5045–5067 (2013)
10. Zhou, A., Qu, B.Y., Li, H., Zhao, S.Z., Suganthan, P.N., Zhang, Q.: Multiobjective evolutionary algorithms: a survey of the state of the art. Swarm Evol. Comput. **1**(1), 32–49 (2011)
11. Zitzler, E., Laumanns, M., Thiele, L.: SPEA2: improving the strength pareto evolutionary algorithm. Eidgenössische Technische Hochschule Zürich (ETH), Institut für Technische Informatik und Kommunikationsnetze (TIK) (2001)

Increasing Data Center Energy Efficiency via Simulation and Optimization of Cooling Circuits - A Practical Approach

Torsten Wilde[1,2](✉), Tanja Clees[3](✉), Hayk Shoukourian[1,2], Nils Hornung[3], Michael Schnell[3], Inna Torgovitskaia[3], Eric Lluch Alvarez[3], Detlef Labrenz[1], and Horst Schwichtenberg[3]

[1] Leibniz Supercomputing Centre of the Bavarian Academy of Science and Humanity, Garching bei München, Germany
torsten.wilde@lrz.de
[2] Technical University Munich (TUM), Munich, Germany
[3] Fraunhofer SCAI (SCAI), Sankt Augustin, Germany
tanja.clees@scai.fraunhofer.de

Abstract. The steady rise in energy consumption by data centers world wide over the last decade and the future 20 MW exascale-challenge in High Performance Computing (HPC) makes saving energy an important consideration for HPC data centers. A move from air-cooled HPC systems to indirect or direct water-cooled systems allowed for the use of chiller-less cold or hot water cooling. However, controlling such systems needs special attention in order to arrive at an optimal compromise of low energy consumption and robust operating conditions. This paper highlights a newly developed concept along with software tools for modeling the data center cooling circuits, collecting data, and simulating and analyzing operating conditions. A first model for the chiller-less cooling loop of the Leibniz Supercomputing Center (LRZ) will be presented and lessons learned will be discussed, demonstrating the possibilities offered by the new concept and tools.

Keywords: HPC · Energy efficiency · Energy reduction · Adsorption · Data center

1 Introduction

For the past decade the energy consumption of data centers has increased substantially. For the year 2014 the world wide power consumption was estimated at 38.84 GW [4]. According to DataCenterDynamics Intelligence [4] the average data center Power Usage Effectivness (PUE) [1] world wide was in the range of 1.81 to 2.0 for 2013. An Uptime Institute survey [15] showed an average PUE of 1.65 for 2013. Using the average of both surveys (1.78, which is most likely very optimistic), 17 GW of total data center power consumption was spent on the infrastructure needed to run and cool the IT systems. Therefore, the reduction

© Springer International Publishing Switzerland 2015
S. Gottwalt et al. (Eds.): EI 2015, LNCS 9424, pp. 208–221, 2015.
DOI: 10.1007/978-3-319-25876-8_18

of energy consumption and the possible reuse of waste heat becomes increasingly important issues in the design and operation of large data centers. When talking about data center energy efficiency one needs to talk about cooling technologies. In the past this was not a strong focus since all data centers used air cooling. Today, many super-computing data centers use a combination of different cooling technologies but there exists a lack of knowledge regarding the energy efficiency of the different cooling infrastructures and their efficient operation.

One specific challenge that HPC data centers are facing is the increasing power variability of the new generation HPC systems. Figure 1 shows the LRZ power profiles for 5 days in Jan. 2014. The power profile of the installed super-computer SuperMUC (Nr.14 Top500 List Nov 2014 [16]) [*P(SuperMUC_IT)* 2nd line from top] shows a lot of variability when compared to the power profile of all other installed IT equipment [*P(Cluster_IT)* 4th dotted line from top] which shows a more traditional power profile, a relatively even line. One of the main reasons for this increased variability is the improvement of the energy efficiency of the new generation CPU's where the power consumption varies strongly with the processor work load [6].

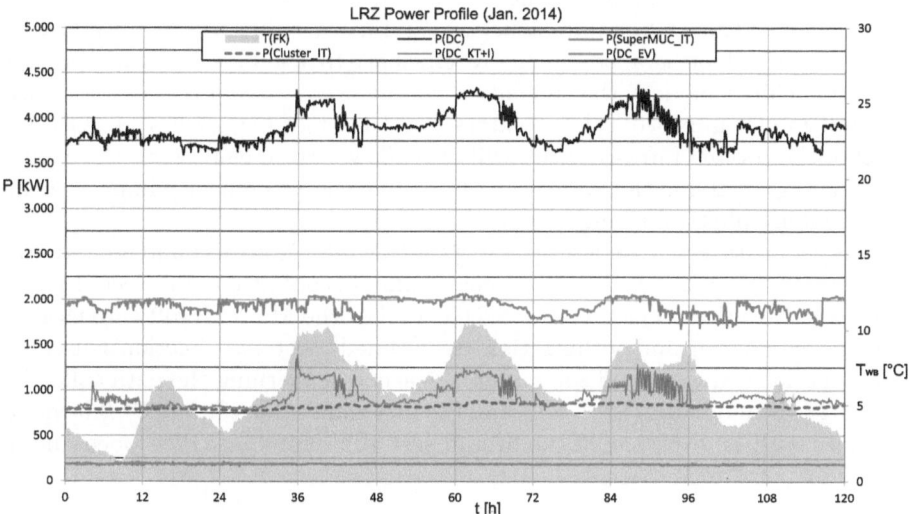

Fig. 1. LRZ data center power consumption Jan 2014.

Current data center cooling infrastructures are still designed based on a constant work load at the maximum power load of the data center (10 MW at LRZ) which does not present the real operating conditions. Figure 1 also highlights the influence of the cooling infrastructure on the data center power consumption. The LRZ data center [8] power profile [*P(DC)* 1st line] shows the variability of SuperMUC [*P(DC_IT)* 2nd line from top] but there are some additional peaks and swings that do not come from the super computer. It doesn't come from the electrical distribution and conversion [$P(DC\_EV)$ 5th line from top] either. As can be seen it comes from the power consumption of the cooling infrastructure [$P(DC\_KT+I)$ 3rd line from top]. There is clearly some relation with the wet-bulb temperature [$T(FK)$ marked area at bottom] which influences the efficiency of the cooling infrastructure.

The question to answer is: why does the cooling infrastructure behave like this, and how can the power spikes and power oscillations of the cooling infrastructure be avoided in order to be more power and energy efficient.

There is a real need to understand the efficient operation and usage of the new cooling technologies, how they differ from traditional air cooled data center operation, how they should be controlled (also in light of high load variability of the new generation HPC systems), and how the design of data centers need to change to make efficient use of the new cooling technologies and possible re-use of heat generated by the HPC systems. The SIMOPEK (Simulation and Optimization of Energy Circuits in Data Centers) project [13], funded by the German Federal Ministry for Education and Research, is investigating the possibility of improving the energy efficiency of the chiller-less cooling loop at the Leibniz Supercomputing Center (LRZ) for specific real world operating points. This will be done by creating a model of the cooling infrastructure, using it to simulate the behavior of the cooling infrastructure, and using multi-criteria optimization to see how the loop can be run more efficient. In combination with possible adsorption chiller models, the simulation can be used to see if the use of adsorption chillers can improve a data centers Total Cost of Ownership (TCO). The model of the cooling infrastructure will also allow virtual changes to the cooling infrastructure enabling the evaluation of possible energy efficiency improvements before making physical changes to it.

This paper starts with a description of the SIMOPEK project and related work (Sect. 2). In particular, the LRZ data center and its cooling circuit (Sect. 2.1), the data collection infrastructure (Sect. 2.2) as well as the MYNTS software used for modeling, simulation and optimization are outlined (Sect. 2.3). Section 3 discusses already performed work and some lessons learned concerning the data collection (Sect. 3.2), the modeling (Sect. 3.1), and the first simulation (Sect. 3.3). The paper concludes with an overview of the overall system architecture developed and an outlook on next steps (Sect. 4).

2   Background

The "4 Pillar Framework for Energy Efficient HPC Data Centers" [19] provides a representation of all parts of a data center that influence its energy efficiency. The four pillars are: Building Infrastructure (Pillar1), HPC System Hardware (Pillar2), HPC System Software (Pillar3), and HPC applications (Pillar4). Figure 2 shows the coverage area of the SIMOPEK project as well as the high level focus areas of each pillar to improve the data center energy efficiency. Currently, data centers focus on each pillar individually. But there are limitations to localized optimization. For example, the cooling infrastructure

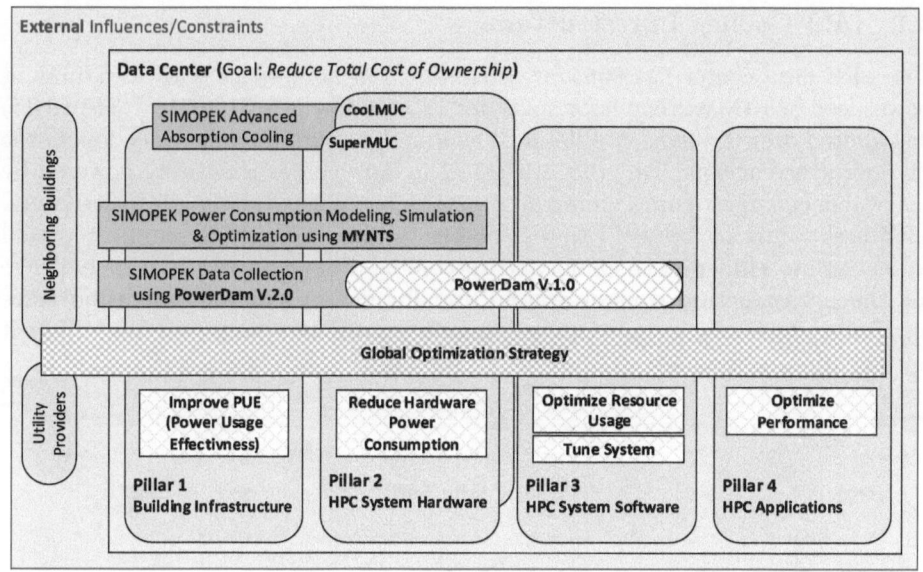

Fig. 2. The four pillar framework for energy efficient HPC data centers and SIMOPEK project coverage.

automation system operates based on static control parameters. It collects current sensor data and adjusts its operation accordingly. But it can't look into the future. It can't distinguish a temporary power consumption spike, where it wouldn't need to adjust, from a relevant increase of the IT system power consumption. Thus it adds more cooling capacity even so the spike might last only a couple of minutes and wouldn't tax the infrastructure resulting in an unnecessary increase in the infrastructure power consumption. But in order to avoid this behavior a meta feedback and control system is required that combines information from Pillar2 (current and future load behavior of the HPC system), Pillar1 (the cooling infrastructure behavior under those conditions), and external influences (temperature and humidity). Another challenge is to decide what the cost-optimal cooling water temperature would be with changing outside conditions. For example, at LRZ, the cooling infrastructure for SuperMUC works towards a set cooling-inlet temperature (40°C summer and 30°C winter) but it might be more efficient to let it vary depending on outside conditions and heat re-use requirements (colder during the nights than during the day, and warmer in summer for hours where the energy usage to cool down to 40°C might be very high because of outside conditions). In the end a global optimization strategy, combining data from all aspects of the data center, might provide the optimal approach to improving the data center energy efficiency. SIMOPEK moves towards that goal connecting Pillar1 and Pillar2 since it is required to be able to model and simulate how the cooling infrastructure can work as efficiently as possible in connection with the load variability of the new generation HPC systems.

2.1   LRZ Cooling Infrastructure

The LRZ data center has $9554\,\mathrm{m}^2$ $(102838\,\mathrm{ft}^2)$ of floor space with a redundant power feed of 10 MW. The floor space for IT equipment is $3160.5\,\mathrm{m}^2$ $(34\,019\,\mathrm{ft}^2)$ distributed over 6 rooms on 3 floors. The infrastructure floor space is double the IT floor space at $6393.5\,\mathrm{m}^2$ $(68\,819\,\mathrm{ft}^2)$. The data center is entirely powered by renewable energy. Figure 3 shows an overview of the LRZ cooling infrastructure. LRZ uses a mix of free cold water cooling (well water), chiller supported cold water cooling, chiller-less cooling, and air cooling. This represents a real challenge for the infrastructure control system and for energy efficient operation since all cooling technologies need to operate in accord but each requires different optimization techniques.

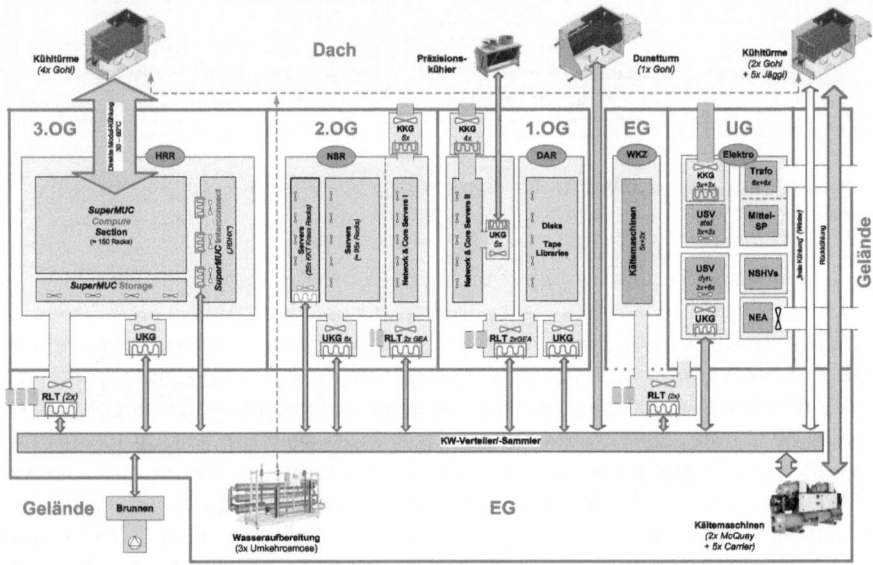

Fig. 3. LRZ cooling infrastructure overview schematic.

Figure 4 shows the chiller-less cooling circuit in more detail. As can be seen there are two separate loops that cool different parts of the data center (HRR and NSR) and both connect to one water distribution. This water distribution is connected to four roof cooling towers (KLT11 - KLT14). At first glance this circuit looks simple enough, but operating the cooling circuit efficiently is not an easy task - The devil is in the details.

For example, an operational challenge is the running of the four roof cooling circuits. Each cooling tower is specified to remove 2 MW of heat. Each cooling circuit feeds from the water distribution, goes over a hydraulic gate (left over from chiller supported cooling, allowing for constant pump speed required by chillers), and into one heat exchanger. The heat exchanger separates the internal

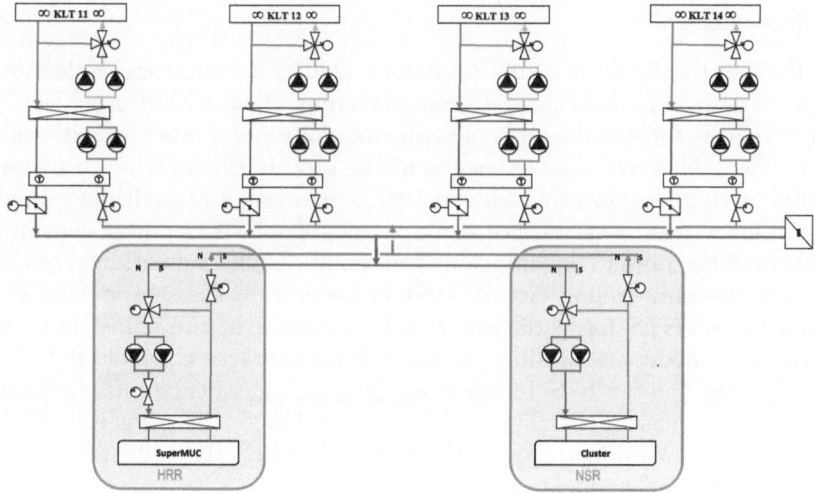

Fig. 4. LRZ chiller-less cooling circuit overview schematic.

water loop (normal water with some additives) from the cooling tower water loop (water-glycol mix). The hydraulic gates and the heat exchanger are specified for a defined temperature spread and a specific pump speed. But since the work load (produced heat) can vary substantially, some load scenarios fall into very inefficient operation points of the cooling circuit. For example, low temperature differences and low pump speed can cause water mixing in the hydraulic gates reducing the real world efficiency of the chiller-less cooling system and leading to a substantial increase in the power consumption of the cooling system.

Besides optimizing the cooling infrastructure operation, the re-use of waste heat can help data centers to avoid or recuperate some costs. LRZ has some preliminary experiences in heat re-use using a SorTech ACS-08 adsorption chiller connected to the PRACE 1IP-WP9 CoolMUC prototype HPC cluster [9]. The cluster is completely room neutral, meaning that there is no requirement for computer room air conditioning (CRAC) units. CoolMUC is arranged in 5 racks. The compute hardware is contained in three racks while the cooling components are contained in two other racks. The connected SorTech ACS-08 adsorption chiller is used to turn the hot water emitted from the cluster into cold water. The cold water is then used to cool the rear-door heat exchanger of a sixth rack. Adsorption chillers come from the solar industry where they are used to generate cold water in combination with solar energy using high input water temperature (70°C and higher). A CoolMUC and adsorption machine model could help to support the notion that adsorption chillers could be used in data centers efficiently, help to determine what type of adsorbent needs to be used to support lower driving temperatures, and help to define how the chiller and the IT system needs to be operated.

2.2   PowerDAM

PowerDAM [11] (Fig. 5) is a tool developed at LRZ to allow the collection and analysis of data from different data center systems. PowerDAM provides a plug-in infrastructure for system data collection and for reports over the collected data allowing for a relatively easy extension of the tool. It also provides a framework for defining virtual sensors which can be a combination of multiple physical as well as other vitrual sensors. For example, the CooLMUC power consumption is the sum of the power consumption of all nodes, of the networking equipment, and of the internal cooling circuit. Besides generating the sensor data xml file required by MYNTS for verification and validation of the chiller-less cooling infrastructure model and feeding the final simulation, the collected data is used to calculate the Energy-to-Solution of applications, to calculate data center Key Performance Indicators (KPI's) such as Power Usage Effectiveness (PUE) [1] and Data center Workload Power Efficiency (DWPE) [18], and for power and energy prediction of applications [12].

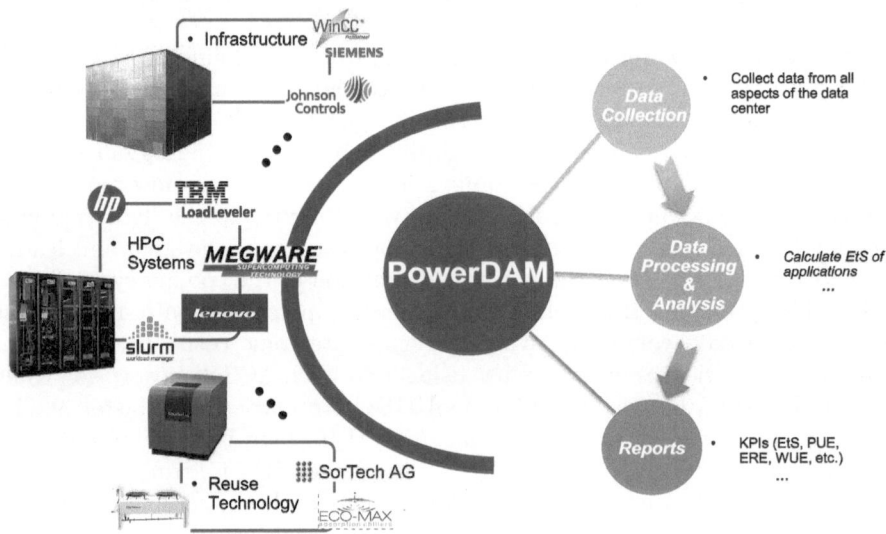

Fig. 5. PowerDAM high level overview.

2.3   MYNTS Software

The MYNTS software package (Multiphysical Network Simulator, cf. [2,5]) was developed for modeling and simulating, as well as analysing and optimizing electric, water, and gas networks. It is used, for instance, by Open Grid Europe, a company running Germany's largest long-distance gas transport network.

MYNTS is built on the fact that such networks can be modeled in a very similar fashion as systems of differential-algebraic equations. The respective non-linear system to be solved - simulated or optimized - is set up based on analytic formulations. MYNTS' NL-converter creates all data necessary for runs with the underlying solver for NLPs (nonlinear problems), namely ipopt [7]. MYNTS allows for user-defined elements (custom physical formulations and constraints).

In combination with net 'O'graph and DesParO [3], graph analysis, statistical analysis, calibration, and robust optimization tasks are supported. The graph analysis features include graph reduction, graph matching, analysis of supply-demand scenarios, and network (de)composition as well as automatic layout methods. DesParO provides metamodeling (efficient interpolation by means of response surfaces with adaptive model build-up), statistical analysis of parameter-criteria dependencies (including own nonlinear correlation, tolerance and quantile estimators, cf. e.g. [10]), and Pareto optimization (multi-objective optimization).

3   First Experiences and Lessons-Learned

This section discusses experiences and lessons learned for the work performed leading to the first cooling circuit model for the chiller-less cooling loop at LRZ.

3.1   Model Creation

In order to set up a model which can be used for analysis, simulation and optimization of the infrastructure network (Pillar1 Fig. 2), the following components are necessary: the layout of the network (topology); the technical description of the network components (elements); the description of inputs, outputs, control logics etc. (scenarios); and the definition of optimization goals.

Ideally, the topology can be obtained or reconstructed by data (e.g. HTML, XML, OPC UA) used in an infrastructure/building management system. Quite often, though, only CAD (computer aided design) data is available for the specific circuits considered. This is also the case for LRZ. Hence, the network topology was setup using a custom semi-automatic procedure: in the first step, the CAD file describing LRZ cooing infrastructure was converted by means of a self-written parser. A major challenge was to reconstruct pipes, technical components, and their connections since the CAD file used dashed lines, non-connected simple sketches etc. In the second step, the resulting topological information was corrected and extended by means of analysis routines and manual work.

To model the physical behavior of the network components, data contained in technical documentation is required. For LRZ, the important components include regulated or unregulated pumps, heat exchangers, cooling towers, hydraulic separators, additional resistors, regulators, one- or three-way valves, and pipes. A special focus in SIMOPEK is on adsorption chillers [17]. A detailed physical description as well as appropriate characteristic maps are available due to a close collaboration with SorTech AG [14]. For use in MYNTS the data was converted to digital form. Since the device names could not be read automatically from the infrastructure plan or management system, a mapping from model generated device names to original device names was generated semi-automatically.

Technical specifications of decisive parameters, settings, and characteristic maps were obtained by digitalizing respective information from the device manuals and from web pages.

Measurement data as well as descriptions of control logics are used: for defining scenarios; to define boundary conditions for setting up several valid simulation models describing important events from the past; to validate simulation results (historical data for offline, online data for dynamic validation); and as input for calibration processes. The latter is necessary if a decisive device acts as a "black box" and cannot be described based on technical documentation. This can happen if the model/type or specifications of the device are not known, or if it is a complex system. Measurement data in its "surroundings" might be used to train approximation models instead of real characteristic maps. DesParO (Sect. 2.3) can be used, for instance. For the LRZ infrastructure, for example, some heat exchangers and the SuperMUC are treated as "black box".

Usually, optimizing a complex infrastructure in combination with possible conflicting operating requirements leads to a multi-objective optimization. The specific optimization selection is not unique and is strongly influenced by goals of the data center and other stakeholders. A traditional approach is to weight goals against each other to form a single optimization criterion. An alternative approach, used in SIMOPEK, is multi-objective optimization in order to compute Pareto-optimal solutions, i.e. best compromises.

Lessons Learned #1: Since infrastructure description data in CAD format doesn't provide enough information for automatic processing, data center need to require parsable topology maps (network topology description and technical details on its elements) of the data center infrastructure either during procurement, or if possible by ordering technical specifications in a suitable electronic format, e.g. tables defining characteristic maps after installation. This will reduce any effort requiring the infrastructure topology substantially. In addition, one should check whether components are really installed according to specifications in the infrastructure plan.

3.2   Collecting and Mapping Data

In most cases, in order to collect the data required for the modelling and simulation of a data center cooling infrastructure, multiple separate data center systems need to be accessed from Pillar1, Pillar2, and Pillar3 (Fig. 2). For the LRZ data center this includes: Pillar1 (JCI - the infrastructure automation system from Johnson Controls, and WinCC - the power distribution monitoring system from SIEMENS); Pillar2 (the relevant information of the IT systems that use the chiller-less cooling circuit - mainly SuperMUC and CooLMUC, and re-use technology information - SorTech Adsorption chiller connected to the CooLMUC, CooLMUC internal cooling circuit information, and CooLMUC power consumption); and Pillar3 (the relevant information of the IT scheduling systems - LoadLeveler and Slurm).

Collecting data from different unrelated systems provided a multitude of challenges.

Since all systems are stand alone systems the time synchronisation between the different system measurements becomes very important. For example, some systems use Coordinated Universal Time (UTC) [SuperMUC PDU data, JCI], others use local time [WinCC, LoadLeveler, Slurm, CoolMUC].

Since PowerDAM uses a specific sensor name schema for its API (Eq. 1) the sensor names from the different systems need to be converted. This needs to be done automatically since systems could have thousands of sensors.

$$RootResource(.Resource) * \_SensorType = Value; Timestamp \qquad (1)$$

On systems that already use an hierarchical naming schema the mapping to the PowerDAM API names is relatively straight forward. For systems, that only ensures the uniqueness of a sensor name (for example, WinCC), the names can be mapped to a flat hierarchy for PowerDAM.

JCI has a name schema for its sensors that seemed well thought out. There is a part identifying the circuit and one that identifies the location. The first Power-DAM mapping was based on the circuit name. Unfortunately, it turned out that a very small number of sensors used the circuit+location as a unique identifier (R3OKLT72 - Location: R3O, Circuit KLT72; R2OKLT72 - Location: R2O, Circuit KLT72, but not the same circuit KLT72). Also there were other naming inconsistencies that required special name handling, making the final solution more complicated than it should have been. (R2OSS_71EZ__PEMW04 follows the naming schema, maps to: R2OSS71.EZ04_MW but @JCSQL:...R2OSS_71EZ__PEMW04 doesn't adhere to JCI internal naming schema mapping to: R2OSS71.EZ04_MW as well).

Lessons Learned #2: Data centers need to be involved in defining the sensor naming schemas for their monitoring systems so that the collected data can be automatically processed by other data center tools. Also it would be beneficial to have tools that can ensure sensor name coherence.

Another unforseen complication was the use of special German characters for two sensor names in WinCC. Since PowerDAM is written using Python 2.7, all internal string conversion is done using ASCII encoding by default. ASCII is fine as long one adheres to the English alphabet. But the German Umlaute requires the use of UTF-8 encoding. This resulted in a partial code re-write because all file processing, XML processing, pattern matching, and string handling needed to use UTF-8 explicitly. This is not required for Python 3.0 (uses UTF-8 as default) but at the time PowerDAM was developed some libraries where not available for Python 3.0 (MySQL-python package still doesn't provide Python 3.0 support as of now).

Lessons Learned #3: If possible, stick to the English alphabet everywhere in your data center software, one will have less worries when processing the data later on. Otherwise be aware of possible UTF-8 related issues.

From an ease-of-access perspective, the IT related system information (Super-MUC, CoolMUC, LoadLeveler and Slurm schedulers) were very easy to access (Linux tool chain). JCI was somewhat harder since the vendor does not consider their tool as providing information to third parties. Also, the design of the JCI tool makes archive sensor data and online sensor data compete for bandwidth. This meant that not all sensor data visualized in the JCI GUI were in the storage data base. Also, event information (pump on/off) was stored in a file not accessible from the outside. The hardest system to access was WinCC. Here the database is encrypted requiring extra tools creating a complicated dependency hierarchy. Everything concerning WinCC is Windows based and accessing the data from Linux is challenging at best. Currently, WinCC data is processed offline (data for a month is exported to csv using the WinCC GUI and the csv file is than imported into PowerDAM).

Lessons Learned #4: Procure data center monitoring and automation tools that allow for easy access to collected data via standards not proprietary ways. If this is not possible, include in the procurement any extra tools and installations needed to access the data.

3.3 Simulation, Analysis and Optimization

The complete preliminary model of the hot-water, chiller-less part of the cooling circuit consists of approx. 800 nodes, 600 pipes, and 250 devices. Current focus is on the NSR cooling circuit (Fig. 4) where CooLMUC is connected. Figure 6 shows, as an example, the cut-out of the NSR cooling circuit from the complete network model. The underlaying graph for the specific circuit was analysed for inconsistencies. The initial simulation results, using real measurement, were used to calibrate the model.

Forming network models means bringing together all components mentioned in Sect. 3.1. Here, a major challenge is the connection of the network elements and the sensor tags. Although SCADA (supervisory control and data acquisition) systems are available, this is a nontrivial technical task.

Lessons Learned #5: Access to the topological data used in the SCADA system(s) employed is mandatory to avoid manual work on mapping tags to

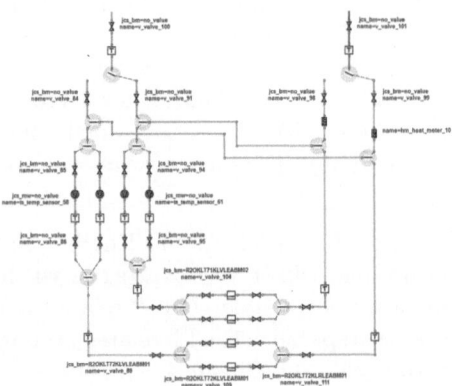

Fig. 6. Exemplary network topology considered: NSR cooling circuit.

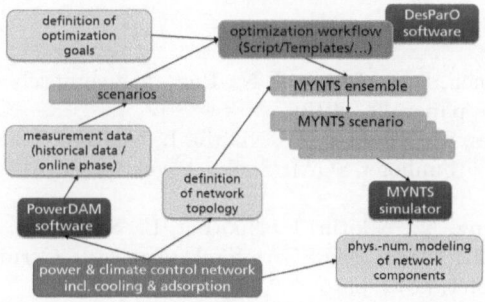

Fig. 7. The overall architecture of the system developed.

elements. One should order either appropriate technical interfaces or directly the respective mapping. Again, this will reduce the effort- and time-to-solution substantially.

The correctness of the collected sensor data is not guaranteed by all systems (some mark invalid data, others don't). Most data problems will become visible during model calibration. Currently, the possibility of adding validation methods for sensor data to PowerDAM are considered. Unfortunately, this is a non-rival task but it would be very helpful if problematic data can be identified and marked before hand.

Lessons Learned #6: Treat collected sensor data with an appropriate amount of scepticism. In most cases, invalid sensor data is the cause of strange data center infrastructure behavior.

4   Summary and Future Work

The overall concept for targeting cooling circuits is depicted in Fig. 7 by means of a high-level description of its components and their connections.

As of now the model of the CoolMUC internal cooling circuit and the supporting data center cooling circuit are developed and tested. The next step is the successful simulation of the model and its use to find an optimal adsorption chiller design. After this, one complete cooling circuit (HRR+NSR and KLT11, see Fig. 4) will be modeled and simulated. The hope is that the simulation can provide information about possible optimization potential. One very debated topic at LRZ, for example, is the need for the hydraulic gates in the chiller-less cooling circuits.

Acknowledgments. The authors would like to thank Jeanette Wilde (LRZ) for her valuable comments and support.

The work presented here has been carried out within the SIMOPEK project [13], which has received funding from the German Federal Ministry for Education and Research under grant number 01IH13007A, at the Leibniz Supercomputing Centre (LRZ) with support of the State of Bavaria, Germany, and the Gauss Centre for Supercomputing (GCS).

References

1. Azevedo, D., French, D.A., Power, E.N.: Pue^{TM}: a comprehensive examination of the metric. White paper 49 (2012)
2. Cassirer, K., Clees, T., Klaassen, B., Nikitin, I., Nikitina, L.: MYNTS User's Manual, Release 3.3. Fraunhofer SCAI, Sankt Augustin (2015). www.scai.fraunhofer.de/mynts
3. Clees, T., Hornung, N., Nikitin, I., Nikitina, L., Steffes-lai, D.: DesParO User's Manual, Release 2.4. Fraunhofer SCAI, Sankt Augustin, Germany, December 2014. www.scai.fraunhofer.de/desparo
4. DCD Intelligence: Is the industry getting better at using power? Data Center Dynamics FOCUS 33, January/February 2014 33, 16–17 (2014). http://content.yudu.com/Library/A2nvau/FocusVolume3issue33/resources/index.htm?referrerUrl=
5. Grundel, S., Hornung, N., Klaassen, B., Benner, P., Clees, T.: Computing surrogates for gas network simulation using model order reduction. In: Koziel, S., Leifsson, L. (eds.) Surrogate-Based Modeling and Optimization: Applications in Engineering. Springer, New York (2013)
6. Hackenberg, D., Schöne, R., Ilsche, T., Molka, D., Schuchart, J., Geyer, R.: An energy efficiency feature survey of the intel haswell processor (2015)
7. Ipopt. http://www.coin-or.org/Ipopt/documentation/
8. Leibniz Supercomputing Centre. http://www.lrz.de
9. Johnsson, L., Netzer, G.: SNIC/KTH; Eric Boyer, CINES; Paul Carpenter, BSC; Radosław Januszewski, PSNC; Giannis Koutsou, CaSToRC; Ole Widar Saastad, SIGMA/UiO; Giannos Stylianou, CaSToRC; Torsten Wilde, LRZ : D9.3.4 Final Report on Prototype Evaluation. PRACE 1IP-WP9 public deliverable p. 44 (2013). http://www.prace-ri.eu/IMG/pdf/d9.3.4_1ip.pdf
10. Rhein, B., Clees, T., Ruschitzka, M.: Robustness measures and numerical approximation of the cumulative density function of response surfaces. Comm. Statistics - Sim. Comp. **43**(1), 1–17 (2014)
11. Shoukourian, H., Wilde, T., Auweter, A., Bode, A.: Monitoring power data: a first step towards a unified energy efficiency evaluation toolset for HPC data centers. Environ. Model. Softw. **56**, 13–26 (2014). http://www.sciencedirect.com/science/article/pii/S1364815213002934, thematic issue on Modelling and evaluating the sustainability of smart solutions
12. Shoukourian, H., Wilde, T., Auweter, A., Bode, A.: Predicting the energy and power consumption of strong and weak scaling HPC applications. Supercomputing Frontiers and Innovations **1**(2), 20–41 (2014)
13. SIMOPEK. http://www.simopek.de
14. SorTech AG. http://www.sortech.de/en/
15. Stansberry, M.: 2013 uptime institute annual data center industry survey report and full results. http://www.data-central.org/resource/collection/BC649AE0-4EDE-92C7-29A659EF0900/uptime-institute-2013-data-center-survey.pdf
16. Top500 List. http://www.top500.org/
17. What is adsoprtion cooling? http://www.sortech.de/en/technology/adsorption/

18. Wilde, T., Auweter, A., Patterson, M., Shoukourian, H., Huber, H., Bode, A., Labrenz, D., Cavazzoni, C.: DWPE, a new data center energy-efficiency metric bridging the gap between infrastructure and workload. In: 2014 International Conference on High Performance Computing Simulation (HPCS), pp. 893–901, July 2014
19. Wilde, T., Auweter, A., Shoukourian, H.: The 4 Pillar Framework for energy efficient HPC data centers. In: Computer Science - Research and Development, pp. 1–11 (2013). http://dx.doi.org/10.1007/s00450-013-0244-6

Author Index